T0173645

# QUANTUM THEORY CANNOT HURT YOU

Marcus Chown is an award-winning writer and broadcaster. Formerly a radio astronomer at the California Institute of Technology in Pasadena, he is currently cosmology consultant of the weekly science magazine *New Scientist*. He is the author of the bestselling *We Need To Talk About Kelvin*, *The Never-Ending Days of Being Dead* and *The Magic Furnace*. He also wrote *The Solar System*, the bestselling app for iPad, which won the Future Book Award 2011. His most recent book is *What a Wonderful World: Life, the Universe and Everything in a Nutshell*.

# QUANTUM THEORY CANNOT HURT YOU

## Understanding the Mind-Blowing
## Building Blocks of the Universe

### Marcus Chown

FABER & FABER

First published in the United States in 2006
by Joseph Henry Press, 500 Fifth Street, NW,
Washington, DC, 20001

First published in the United Kingdom in 2007
by Faber & Faber Limited
The Bindery, 51 Hatton Garden
London EC1N 8HN

Printed in the UK by CPI Group (UK) Ltd, Croydon, CR0 4YY

A CIP record for this book
is available from the British Library

ISBN 978–0–571–31502–4

24

To Patrick, who, when I'm down and wondering why everyone seems to be against me, consoles me by saying: "That's because you're a complete bastard, Marcus!"

# Contents

# FOREWORD

One of the following is true:

- Every breath you take contains an atom breathed out by Marilyn Monroe.
- There is a liquid that can run uphill.
- You age faster at the top of a building than at the bottom.
- An atom can be in many different places at once, the equivalent of you being in New York and London at the same time.
- The entire human race would fit in the volume of a sugar cube.
- One per cent of the static on a television tuned between stations is the relic of the Big Bang.
- Time travel is not forbidden by the laws of physics.
- A cup of coffee weighs more when it is hot than when it is cold.
- The faster you travel, the slimmer you get.

No, I'm joking. They are all true!

As a science writer I am constantly amazed by how much stranger science is than science fiction, how much more incredible the Universe is than anything we could possibly have invented. Despite this, however, very few of the extraordinary discoveries of the past century seem to have trickled through into the public consciousness.

The two towering achievements of the past 100 years are quantum theory, our picture of atoms and their constituents, and Einstein's general theory of relativity, our picture of space, time, and gravity. Between them the two explain virtually everything about the world and about us. In fact, it can be argued that quantum theory has actually *created* the modern world, not only explaining why the ground beneath our feet is solid and why the Sun shines but also making possible computers and lasers and nuclear reactors. Relativity may not be as ubiquitous in the everyday world. Nevertheless, it has taught us that there are things called black holes from which nothing, not even light, can escape; that the Universe has not existed forever but was born in a titanic explosion called the Big Bang; and that time machines—remarkably—may be possible.

Although I have read many popular accounts of these topics, the explanations have often left me baffled, even with my science background. I can only guess what it must be like for nonscientists.

Einstein said: "Most of the fundamental ideas of science are essentially simple and may, as a rule, be expressed in a language comprehensible to everyone." All my experience tells me he was right. My idea in writing this book was to try to help ordinary people understand the principal ideas of 21st-century physics. All I had to do was identify the key ideas behind quantum theory and relativity, which turn out to be deceptively simple, and then show how absolutely everything else follows from them logically and unavoidably.

Easier said than done. Quantum theory in particular is a patchwork of fragments, accrued over the past 80 years, that nobody seems to have sewn together into a seamless garment. What's more, crucial pieces of the theory, such as "decoherence"—which explains why atoms but not people can be in two places at once—seem to be beyond the power of physicists to communicate in any intelligible way. After corresponding with many "experts," and beginning to think that decoherence should be renamed "incoherence," it dawned on me that maybe the experts didn't completely understand it themselves. In a way this was liberating. Since a coherent picture seemed not to exist, I realised that I had to piece together my own from insights gleaned from different people. Because of this, many of the explanations given here you will not find anywhere else. I hope they help lift some of the fog that surrounds the key ideas of modern physics and that you can begin to appreciate what a breathtakingly amazing Universe we find ourselves in.

# Acknowledgements

My thanks to the following people who either helped me directly, inspired me, or simply encouraged me during the writing of this book: My dad, Karen, Sara Menguc, Jeffrey Robbins, Neil Belton, Henry Volans, Rachel Marcus, Moses Cardona, Brian Clegg, Professor Tony Hey, Kate Oldfield, Vivien James, Brian May, Dr. Bruce Bassett, Dr. Larry Schulman, Dr. Wojciech Zurek, Sir Martin Rees, Allison Chown, Colin Wellman, Rosie and Tim Chown, Patrick O'Halloran, Julie and Dave Mayes, Stephen Hedges, Sue O'Malley, Sarah Topalian, Dr. David Deutsch, Alexandra Feacham, Nick Mayhew-Smith, Elisabeth Geake, Al Jones, David Hough, Fred Barnum, Pam Young, Roy Perry, Hazel Muir, Stuart and Nikki Clark, Simon Ings, Barry Fox, Spencer Bright, Karen Gunnell, Jo Gunnell, Pat and Brian Chilver, Stella Barlow, Silvano Mazzon, Barbara Pell and David, Julia Bateson, Anne Ursell, Barbara Kiser, Dottie Friedli, Jon Holland, Martin Dollard, Sylvia and Sarah Kefyalew, Matilda and Dennis and Amanda and Andrew Buckley, Diane and Peter and Ciaran and Lucy Tomlin, Eric Gourlay, Paul Brandford. It goes without saying, I hope, that none of these people are responsible for any errors.

# PART ONE

# SMALL THINGS

# 1

# Breathing in Einstein

How we discovered that everything is made of
atoms and that atoms are mostly empty space

*A hydrogen atom in a cell at the end of my nose was once part of an
elephant's trunk.*

Jostein Gaarder

*We never had any intention of using the weapon. But they were such a
terribly troublesome race. They insisted on seeing us as the "enemy" despite all our efforts at reassurance. When they fired their entire nuclear
stockpile at our ship, orbiting high above their blue planet, our patience
simply ran out.*

*The weapon was simple but effective. It squeezed out all the empty
space from matter.*

*As the commander of our Sirian expedition examined the shim-
mering metallic cube, barely 1 centimetre across, he shook his primary
head despairingly. Hard to believe that this was all that was left of the
"human race"!*

If the idea of the entire human race fitting into the volume of a sugar
cube sounds like science fiction, think again. It is a remarkable fact
that 99.9999999999999 per cent of the volume of ordinary matter is
empty space. If there were some way to squeeze all the empty space
out of the atoms in our bodies, humanity would indeed fit into the
space occupied by a sugar cube.

The appalling emptiness of atoms is only one of the extraordinary characteristics of the building blocks of matter. Another, of course, is their size. It would take 10 million atoms laid end to end to span the width of a single full stop on this page, which raises the question, how did we ever discover that everything is made of atoms in the first place?

The idea that everything is made of atoms was actually first suggested by the Greek philosopher Democritus in about 440 BC.[1] Picking up a rock—or it may have been a branch or a clay pot—he asked himself the question: "If I cut this in half, then in half again, can I go on cutting it in half forever?" His answer was an emphatic *no*. It was inconceivable to him that matter could be subdivided forever. Sooner or later, he reasoned, a tiny grain of matter would be reached that could be cut no smaller. Since the Greek for "uncuttable" was "*a-tomos,*" Democritus called the hypothetical building blocks of all matter "atoms."

Since atoms were too small to be seen with the senses, finding evidence for them was always going to be difficult. Nevertheless, a way was found by the 18th-century Swiss mathematician Daniel Bernoulli. Bernoulli realised that, although atoms were impossible to observe directly, it might still be possible to observe them indirectly. In particular, he reasoned that if a large enough number of atoms acted together, they might have a big enough effect to be obvious in the everyday world. All he needed was to find a place in nature where this happened. He found one—in a "gas."

Bernoulli imagined a gas like air or steam as a collection of billions upon billions of atoms in perpetual frenzied motion like a swarm of angry bees. This vivid picture immediately suggested an explanation for the "pressure" of a gas, which kept a balloon inflated

---

[1]Some of these ideas were covered in my earlier book, *The Magic Furnace* (Vintage, London, 2000). Apologies to those who have read it. In my defense, it is necessary to know some basic things about the atom in order to appreciate the chapters that follow on quantum theory, which is essentially a theory of the atomic world.

or pushed against the piston of a steam engine. When confined in any container, the atoms of a gas would drum relentlessly on the walls like hailstones on a tin roof. Their combined effect would be to create a jittery force that, to our coarse senses, would seem like a constant force pushing back the walls.

But Bernoulli's microscopic explanation of pressure provided more than a convenient mental picture of what was going on in a gas. Crucially, it led to a specific prediction. If a gas were squeezed into half its original volume, the gas atoms would need to fly only half as far between collisions with the container walls. They would therefore collide twice as frequently with those walls, doubling the pressure. And if the gas were squeezed into a third of its volume, the atoms would collide three times as frequently, trebling the pressure. And so on.

Exactly this behaviour was observed by the English scientist Robert Boyle in 1660. It confirmed Bernoulli's picture of a gas. And since Bernoulli's picture was of tiny grainlike atoms flying hither and thither through empty space, it bolstered the case for the existence of atoms. Despite this success, however, definitive evidence for the existence of atoms did not come until the beginning of the 20th century. It was buried in an obscure phenomenon called Brownian motion.

Brownian motion is named after Robert Brown, a botanist who sailed to Australia on the Flinders expedition of 1801. During his time down under, he classified 4,000 species of antipodean plants; in the process, he discovered the nucleus of living cells. But he is best remembered for his observation in 1827 of pollen grains suspended in water. To Brown, squinting through a magnifying lens, it seemed as if the grains were undergoing a curious jittery motion, zigzagging their way through the liquid like drunkards lurching home from a pub.

Brown never solved the mystery of the wayward pollen grains. That breakthrough had to wait for Albert Einstein, aged 26 and in the midst of the greatest explosion of creativity in the history of science. In his "miraculous year" of 1905, not only did Einstein overthrow

Newton, supplanting Newtonian ideas about motion with his special theory of relativity, but he finally penetrated the 80-year-old mystery of Brownian motion.

The reason for the crazy dance of pollen grains, according to Einstein, was that they were under continual machine-gun bombardment by tiny water molecules. Imagine a giant inflatable rubber ball, taller than a person, being pushed about a field by a large number of people. If each person pushes in their own particular direction, without any regard for the others, at any instant there will be slightly more people on one side than another. This imbalance is enough to cause the ball to move erratically about the field. Similarly, the erratic motion of a pollen grain can be caused by slightly more water molecules bombarding it from one side than from another.

Einstein devised a mathematical theory to describe Brownian motion. It predicted how far and how fast the average pollen grain should travel in response to the relentless battering it was receiving from the water molecules all around. Everything hinged on the size of the water molecules, since the bigger they were the bigger would be the imbalance of forces on the pollen grain and the more exaggerated its consequent Brownian motion.

The French physicist Jean Baptiste Perrin compared his observations of water-suspended "gamboge" particles, a yellow gum resin from a Cambodian tree, with the predictions of Einstein's theory. He was able to deduce the size of water molecules and hence the atoms out of which they were built. He concluded that atoms were only about one 10-billionth of a metre across—so small that it would take 10 million, laid end to end, to span the width of a full stop.

Atoms were so small, in fact, that if the billions upon billions of them in a single breath were spread evenly throughout Earth's atmosphere, every breath-sized volume of the atmosphere would end up containing several of those atoms. Put another way, every breath you take contains at least one atom breathed out by Albert Einstein—or Julius Caesar or Marilyn Monroe or even the last Tyrannosaurus Rex to walk on Earth!

What is more, the atoms of Earth's "biosphere" are constantly recycled. When an organism dies, it decays and its constituent atoms are returned to the soil and the atmosphere to be incorporated into plants that are later eaten by animals and humans. "A carbon atom in my cardiac muscle was once in the tail of a dinosaur," writes Norwegian novelist Jostein Gaarder in *Sophie's World*.

Brownian motion was the most powerful evidence for the existence of atoms. Nobody who peered down a microscope and saw the crazy dance of pollen grains under relentless bombardment could doubt that the world was ultimately made from tiny, bulletlike particles. But watching jittery pollen grains—the effect of atoms—was not the same as actually *seeing* atoms. This had to wait until 1980 and the invention of a remarkable device called the scanning tunnelling microscope (STM).

The idea behind the STM, as it became known, was very simple. A blind person can "see" someone's face simply by running a finger over it and building up a picture in their mind. The STM works in a similar way. The difference is that the "finger" is a finger of metal, a tiny stylus reminiscent of an old-fashioned gramophone needle. By dragging the needle across the surface of a material and feeding its up-and-down motion into a computer, it is possible to build up a detailed picture of the undulations of the atomic terrain.[2]

---

[2]Of course, there is no way a needle can actually feel a surface like a human finger can. However, if the needle is charged with electricity and placed extremely close to a conducting surface, a minuscule but measurable electric current leaps the gap between the tip of the needle and the surface. It is known as a "tunnelling current", and it has a crucial property that can be exploited: the size of the current is extraordinarily sensitive to the width of the gap. If the needle is moved even a shade closer to the surface, the current grows very rapidly; if it is pulled away a fraction, the current plummets. The size of the tunnelling current therefore reveals the distance between the needle tip and the surface. It gives the needle an artificial sense of touch.

Of course, there is a bit more to it than that. Although the principle of the invention was simple, there were formidable practical difficulties in its realisation. For instance, a needle had to be found that was fine enough to "feel" atoms. The Nobel Prize committee certainly recognised the difficulties. It awarded Gerd Binnig and Heinrich Rohrer, the IBM researchers behind the STM, the 1986 Nobel Prize for Physics.

Binnig and Rohrer were the first people in history to actually "see" an atom. Their STM images were some of the most remarkable in the history of science, ranking alongside that of Earth rising above the gray desolation of the Moon or the sweeping spiral staircase of DNA. Atoms looked like tiny footballs. They looked like oranges, stacked in boxes, row on row. But most of all they looked like the tiny hard grains of matter that Democritus had seen so clearly in his mind's eye, 2,400 years before. No one else has ever made a prediction that far in advance of experimental confirmation.

But only one side of the atom was revealed by the STM. As Democritus himself had realised, atoms were a lot more than simply tiny grains in ceaseless motion.

## NATURE'S LEGO BRICKS

Atoms are nature's Lego bricks. They come in a variety of different shapes and sizes, and by joining them together in any number of different ways, it is possible to make a rose, a bar of gold, or a human being. Everything is in the combinations.

The American Nobel Prize winner Richard Feynman said: "If in some cataclysm all of scientific knowledge were destroyed and only one sentence passed on to succeeding generations, what statement would convey the most information in the fewest words?" He was in no doubt: "Everything is made of atoms."

The key step in proving that atoms are nature's Lego bricks was identifying the different kinds of atoms. However, the fact that atoms were far too small to be perceived directly by the senses made the task every bit as formidable as proving that atoms were tiny grains of matter in ceaseless motion. The only way to identify different types of atoms was to find substances that were made exclusively out of atoms of a single kind.

In 1789 the French aristocrat Antoine Lavoisier compiled a list of substances that he believed could not, by any means, be broken down into simpler substances. There were 23 "elements" in Lavoisier's list. Though some later turned out not to be elements, many—including gold, silver, iron, and mercury—were indeed elemental. Within 40 years of Lavoisier's death at the guillotine in 1794, the list of elements had grown to include close to 50. Nowadays, we know of 92 naturally occurring elements, from hydrogen, the lightest, to uranium, the heaviest.

But what makes one atom different from another? For instance, how does a hydrogen atom differ from a uranium atom? The answer would come only by probing their internal structures. But atoms are so fantastically small. It seemed impossible that anyone would ever find a way to look inside one. But one man did—a New Zealander named Ernest Rutherford. His ingenious idea was to use atoms to look inside other atoms.

## THE MOTH IN THE CATHEDRAL

The phenomenon that laid bare the structure of atoms was radioactivity, discovered by the French chemist Henri Becquerel in 1896. Between 1901 and 1903, Rutherford and the English chemist Frederick Soddy found strong evidence that a radioactive atom is simply a heavy atom that is seething with excess energy. Inevitably, after a second or a year or a million years, it sheds this surplus energy by expelling some kind of particle at high speed. Physicists say it disintegrates, or "decays," into an atom of a slightly lighter element.

One such decay particle was the alpha particle, which Rutherford and the young German physicist Hans Geiger demonstrated was simply an atom of helium, the second lightest element after hydrogen.

In 1903, Rutherford had measured the speed of alpha particles expelled from atoms of radioactive radium. It was an astonishing 25,000 kilometres per second—100,000 times faster than a present-day passenger jet. Here, Rutherford realised, was a perfect bullet to smash into atoms and reveal what was deep inside.

The idea was simple. Fire alpha particles into an atom. If they encountered anything hard blocking their way, they would be deflected from their path. By firing thousands upon thousands of alpha particles and observing how they were deflected, it would be possible to build up a detailed picture of the interior of an atom.

Rutherford's experiment was carried out in 1909 by Geiger and a young New Zealand physicist called Ernest Marsden. Their "alpha-scattering" experiment used a small sample of radium, which fired off alpha particles like microscopic fireworks. The sample was placed behind a lead screen containing a narrow slit, so a thread-thin stream of alpha particles emerged on the far side. It was the world's smallest machine gun, rattling out microscopic bullets.

In the firing line Geiger and Marsden placed a sheet of gold foil only a few thousand atoms thick. It was insubstantial enough that all the alpha particles from the miniature machine gun would pass through. But it was substantial enough that, during their transit, some would pass close enough to gold atoms to be deflected slightly from their path.

At the time of Geiger and Marsden's experiment, one particle from inside the atom had already been identified. The electron had been discovered by the British physicist "J. J." Thomson in 1895. Ridiculously tiny particles—each about 2,000 times smaller than even a hydrogen atom—had turned out to be the elusive particles of electricity. Ripped free from atoms, they surged along a copper wire amid billions of others, creating an electric current.

The electron was the first subatomic particle. It carried a negative electric charge. Nobody knows exactly what electric charge is, only that it comes in two forms: negative and positive. Ordinary matter, which consists of atoms, has no net electrical charge. In ordinary atoms, then, the negative charge of the electrons is always perfectly balanced by the positive charge of something else. It is a characteristic of electrical charge that unlike charges attract each other whereas like charges repel each other. Consequently, there is a force of attraction between an atom's negatively charged electrons and its positively charged something else. It is this attraction that glues the whole thing together.

Not long after the discovery of the electron, Thomson used these insights to concoct the first-ever scientific picture of the atom. He visualised it as a multitude of tiny electrons embedded "like raisins in a plum pudding" in a diffuse ball of positive charge. It was Thomson's plum pudding model of the atom that Geiger and Marsden expected to confirm with their alpha-scattering experiment.

They were to be disappointed.

The thing that blew the plum pudding model out of the water was a rare but remarkable event. One out of every 8,000 alpha particles fired by the miniature machine gun actually bounced back from the gold foil!

According to Thomson's plum pudding model, an atom consisted of a multitude of pin-prick electrons embedded in a diffuse globe of positive charge. The alpha particles that Geiger and Marsden were firing into this flimsy arrangement, on the other hand, were unstoppable subatomic express trains, each as heavy as around 8,000 electrons. The chance of such a massive particle being wildly deflected from its path was about as great as that of a real express train being derailed by a runaway dolls pram. As Rutherford put it: "It was almost as incredible as if you fired a 15-inch shell at a piece of tissue paper and it came back and hit you!"

Geiger and Marsden's extraordinary result could only mean that an atom was not a flimsy thing at all. Something buried deep inside it could stop a subatomic express train dead in its tracks and turn it around. That something could only be a tiny nugget of positive charge sitting at the dead centre of an atom and repelling the positive charge of an incoming alpha particle. Since the nugget was capable of standing up to a massive alpha particle without being knocked to kingdom come, it too must be massive. In fact, it must contain almost all of the mass of an atom.

Rutherford had discovered the atomic nucleus.

The picture of the interior of the atom that emerged was as unlike Thomson's plum pudding picture as was possible to imagine. It was a miniature solar system in which negatively charged electrons were attracted to the positive charge of the nucleus and orbited it like planets around the Sun. The nucleus had to be at least as massive as an alpha particle—and probably a lot more so—for the nucleus with which it collided not to be kicked out of the atom. It therefore had to contain more than 99.9 per cent of the atom's mass.[3]

The nucleus was very, very tiny. Only if nature crammed a large positive charge into a very small volume could a nucleus exert a repulsive force so overwhelming that it could make an alpha particle execute a U-turn. What was most striking about Rutherford's vision of an atom was, therefore, its appalling emptiness. The playwright Tom Stoppard put it beautifully in his play *Hapgood:* "Now make a fist, and if your fist is as big as the nucleus of an atom then the atom is as big as St Paul's, and if it happens to be a hydrogen atom then it has a single electron flitting about like a moth in an empty cathedral, now by the dome, now by the altar."

Despite its appearance of solidity, the familiar world was actually no more substantial than a ghost. Matter, whether in the form of a chair, a human being, or a star, was almost exclusively empty space.

---

[3]Eventually, physicists would discover that the nucleus contains two particles: positively charged protons and uncharged, or neutral, neutrons. The number of protons in a nucleus is always exactly balanced by an equal number of electrons in orbit about it. The difference between atoms is in the number of protons in their nuclei (and consequently the number of electrons in orbit). For instance, hydrogen has one proton in its nucleus and uranium a whopping 92.

What substance an atom possessed resided in its impossibly small nucleus—100,000 times smaller than a complete atom.

Put another way, matter is spread extremely thinly. If it were possible to squeeze out all the surplus empty space, matter would take up hardly any room at all. In fact, this is perfectly possible. Although an easy way to squeeze the human race down to the size of a sugar cube probably does not exist, a way does exist to squeeze the matter of a massive star into the smallest volume possible. The squeezing is done by tremendously strong gravity, and the result is a neutron star. Such an object packs the enormous mass of a body the size of the Sun into a volume no bigger than Mount Everest.[4]

## THE IMPOSSIBLE ATOM

Rutherford's picture of the atom as a miniature solar system with tiny electrons flitting about a dense atomic nucleus like planets around the Sun was a triumph of experimental science. Unfortunately, it had a slight problem. It was totally incompatible with all known physics!

According to Maxwell's theory of electromagnetism—which described all electrical and magnetic phenomena—whenever a charged particle accelerates, changing its speed or direction of motion, it gives out electromagnetic waves—light. An electron is a charged particle. As it circles a nucleus, it perpetually changes its direction; so it should act like a miniature lighthouse, constantly broadcasting light waves into space. The problem is that this would be a catastrophe for any atom. After all, the energy radiated as light has to come from somewhere, and it can only come from the electron itself. Sapped continually of energy, it should spiral ever closer to the centre of the atom. Calculations showed that it would collide with the nucleus within a mere hundred-millionth of a second. By rights, atoms should not exist.

---

[4]See Chapter 4, "Uncertainty and the Limits of Knowledge."

But atoms do exist. We and the world around us are proof enough of that. Far from expiring in a hundred-millionth of a second, atoms have survived intact since the earliest times of the Universe almost 14 billion years ago. Some crucial ingredient must be missing from Rutherford's picture of the atom. That ingredient is a revolutionary new kind of physics—quantum theory.

# 2

# WHY GOD PLAYS DICE
# WITH THE UNIVERSE

HOW WE DISCOVERED THAT THINGS IN THE WORLD OF ATOMS
HAPPEN FOR NO REASON AT ALL

*A philosopher once said, "It is necessary for the very existence of science that the same conditions always produce the same results." Well, they don't!*

Richard Feynman

*It's 2025 and high on a desolate mountain top a giant 100-metre telescope tracks around the night sky. It locks onto a proto-galaxy at the edge of the observable Universe and feeble light, which has been travelling through space since long before Earth was born, is concentrated by the telescope mirror onto ultrasensitive electronic detectors. Inside the telescope dome, seated at a control panel not unlike the console of the starship* Enterprise, *the astronomers watch a fuzzy image of the galaxy swim into view on a computer monitor. Someone turns up a loudspeaker and a deafening crackle fills the control room. It sounds like machine gun fire; it sounds like rain drumming on a tin roof. In fact, it is the sound of tiny particles of light raining down on the telescope's detectors from the very depths of space.*

To these astronomers, who spend their careers straining to see the weakest sources of light in the Universe, it is a self-evident fact that light is a stream of tiny bulletlike particles—photons. Not long ago, however, the scientific community had to be dragged kicking and screaming to an acceptance of this idea. In fact, it's fair to say that the discovery that light comes in discrete chunks, or quanta, was the single most shocking discovery in the history of science. It swept away the comfort blanket of pre-20th-century science and exposed physicists to the harsh reality of an *Alice in Wonderland* universe where things happen because they happen, with utter disregard for the civilised laws of cause and effect.

The first person to realise that light was made of photons was Einstein. Only by imagining it as a stream of tiny particles could he make sense of á phenomenon known as the photoelectric effect. When you walk into a supermarket and the doors open for you automatically, they are being controlled by the photoelectric effect. Certain metals, when exposed to light, eject particles of electricity—electrons. When incorporated into a photocell, such a metal generates a small electric current as long as a light beam is falling on it. A shopper who breaks the beam chokes off the current, signalling the supermarket doors to swish aside.

One of the many peculiar characteristics of the photoelectric effect is that, even if a very weak light is used, the electrons are kicked out of the metal instantaneously—that is, with no delay whatsoever.[1] This is inexplicable if light is a wave. The reason is that a wave, being a spread-out thing, will interact with a large number of electrons in the metal. Some will inevitably be kicked out after others. In fact, some of

---

[1]Another interesting characteristic of the photoelectric effect is that no electrons at all are emitted by the metal if it is illuminated by light with a wavelength—a measure of the distance between successive wave crests—above a certain threshold. This, as Einstein realised, is because photons of light have an energy that goes down with increasing wavelength. And below a certain wavelength the photons have insufficient energy to kick an electron out of the metal.

the electrons could easily be emitted 10 minutes or so after light is shone on the metal.

So how is it possible that the electrons are kicked out of the metal instantaneously? There is only one way—if each electron is kicked out of the metal by *a single particle of light.*

Even stronger evidence that light consists of tiny bulletlike particles comes from the Compton effect. When electrons are exposed to X-rays—a high-energy kind of light—they recoil in exactly the way they would if they were billiard balls being struck by other billiard balls.

On the surface, the discovery that light behaves like a stream of tiny particles may not appear very remarkable or surprising. But it is. The reason is that there is also abundant and compelling evidence that light is something as different from a stream of particles as it is possible to imagine—a wave.

### RIPPLES ON A SEA OF SPACE

At the beginning of the 19th century, the English physician Thomas Young, famous for decoding the Rosetta stone independently of the Frenchman Jean François Champollion, took an opaque screen, made two vertical slits in it very close together, and shone light of a single colour onto them. If light were a wave, he reasoned, each slit would serve as a new source of waves, which would spread out on the far side of the screen like concentric ripples on a pond.

A characteristic property exhibited by waves is interference. When two similar waves pass through each other, they reinforce each other where the crest of one wave coincides with the crest of another, and they cancel each other out where the crest of one coincides with the trough of the other. Look at a puddle during a rain shower and you will see the ripples from each raindrop spreading out and "constructively" and "destructively" interfering with each other.

In the path of the light emerging from his two slits Young interposed a second, white, screen. He immediately saw a series of alternating dark and light vertical stripes, much like the lines on a super-market bar code. This interference pattern was irrefutable evidence that light was a wave. Where the light ripples from the two slits were in step, matching crest for crest, the light was boosted in brightness; where they were out of step, the light was cancelled out.

Using his "double slit" apparatus, Young was able to determine the wavelength of light. He discovered it was a mere thousandth of a millimetre—far smaller than the width of a human hair—explaining why nobody had guessed light was a wave before.

For the next two centuries, Young's picture of light as ripples on a sea of space reigned supreme in explaining all known phenomena involving light. But by the end of the 19th century, trouble was brewing. Although few people noticed at first, the picture of light as a wave and the picture of the atom as a tiny mote of matter were irreconcilable. The difficulty was at the interface, the place where light meets matter.

## TWO FACES OF A SINGLE COIN

The interaction between light and matter is of crucial importance to the everyday world. If the atoms in the filament of a bulb did not spit out light, we could not illuminate our homes. If the atoms in the retina of your eye did not absorb light, you would be unable to read these words. The trouble is that the emission and absorption of light by atoms are impossible to understand if light is a wave.

An atom is a highly localised thing, confined to a tiny region of space, whereas a light wave is a spread-out thing that fills a large amount of space. So, when light is absorbed by an atom, how does such a big thing manage to squeeze into such a tiny thing? And when light is emitted by an atom, how does such a small thing manage to cough out such a big thing?

Common sense says that the only way light can be absorbed or emitted by a small localised thing is if it too is a small, localised thing. "Nothing fits inside a snake like another snake," as the saying goes.

Light, however, is known to be a wave. The only way out of the conundrum was for physicists to throw up their hands in despair and grudgingly accept that light is both a wave and a particle. But surely something cannot be simultaneously localised and spreadout? In the everyday world, this is perfectly true. Crucially, however, we are not talking about the everyday world; we are talking about the microscopic world.

The microscopic world of atoms and photons turns out to be nothing like the familiar realm of trees and clouds and people. Since it is a domain millions of times smaller than the realm of familiar objects, why should it be? Light really is both a particle and a wave. Or more correctly, light is "something else" for which there is no word in our everyday language and nothing to compare it with in the everyday world. Like a coin with two faces, all we can see are its particle-like face and its wavelike face. What light *actually is* is as unknowable as the colour blue is to a blind man.

Light sometimes behaves like a wave and sometimes like a stream of particles. This was an extremely difficult thing for the physicists of the early 20th century to accept. But they had no choice; it was what nature was telling them. "On Mondays, Wednesdays and Fridays, we teach the wave theory and on Tuesdays, Thursdays and Saturdays the particle theory," joked the English physicist William Bragg in 1921.

Bragg's pragmatism was admirable. Unfortunately, it was not enough to save physics from disaster. As Einstein first realised, the dual wave-particle nature of light was a catastrophe. It was not just impossible to visualise, it was completely incompatible with all physics that had gone before.

## WAVING GOODBYE TO CERTAINTY

Take a window. If you look closely you can see a faint reflection of your face. This is because glass is not perfectly transparent. It transmits about 95 per cent of the light striking it while reflecting the remaining 5 per cent. If light is a wave, this is perfectly easy to understand. The wave simply splits into a big wave that goes through the window and a much smaller wave that comes back. Think of the bow wave from a speedboat. If it encounters a half-submerged piece of driftwood, a large part of the wave continues on its way while a small part doubles back on itself.

But while this behaviour is easy to understand if light is a wave, it is extremely difficult to understand if light is a stream of identical bulletlike particles. After all, if all the photons are identical, it stands to reason that each should be affected by the window in an identical way. Think of David Beckham taking a free kick over and over again. If the soccer balls are identical and he kicks each one in exactly the same way, they will all curl through the air and hit the same spot at the back of goal. It's hard to imagine the majority of the balls peppering the same spot while a minority flies off to the corner flag.

How, then, is it possible that a stream of absolutely identical photons can impinge on a window and 95 per cent can go right through while 5 per cent come back? As Einstein realised, there is only one way: if the word "identical" has a very different meaning in the microscopic world than in the everyday world—a diminished, cut-down meaning.

In the microscopic domain, it turns out, identical things do not behave in identical ways in identical circumstances. Instead, they merely have an identical *chance* of behaving in any particular way. Each individual photon arriving at the window has exactly the same *chance* of being transmitted as any of its fellows—95 per cent—and exactly the same *chance* of being reflected—5 per cent. There is absolutely no way to know for certain what will happen to a given photon. Whether it is transmitted or reflected is entirely down to random chance.

In the early 20th century, this unpredictability was something radically new in the world. Imagine a roulette wheel and a ball jouncing around as the wheel spins. We think of the number the ball comes to rest on when the wheel finally halts as inherently unpredictable. But it is not—not really. If it were possible to know the initial trajectory of the ball, the initial speed of the wheel, the way the air currents changed from instant to instant in the casino, and so on, the laws of physics could be used to predict with 100 per cent certainty where the ball will end up. The same is true with the tossing of a coin. If it were possible to know how much force is applied in the flipping, the exact shape of the coin, and so on, the laws of physics could predict with 100 per cent certainty whether the coin will come down heads or tails.

Nothing in the everyday world is fundamentally unpredictable; nothing is truly random. The reason we cannot predict the outcome of a game of roulette or of the toss of a coin is that there is simply too much information for us to take into account. But in principle—and this is the key point—there is nothing to prevent us from predicting both.

Contrast this with the microscopic world of photons. It matters not the slightest how much information we have in our possession. It is impossible to predict whether a given photon will be transmitted or reflected by a window—even in principle. A roulette ball does what it does for a reason—because of the interplay of myriad subtle forces. A photon does what it does for no reason whatsoever! The unpredictability of the microscopic world is fundamental. It is truly something new under the Sun.

And what is true of photons turns out to be true of all the denizens of the microscopic realm. A bomb detonates because its timer tells it to or because a vibration disturbs it or because its chemicals have suddenly become degraded. An unstable, or "radioactive," atom simply detonates. There is absolutely no discernible difference between one that detonates at this moment and an identical atom that waits quietly for 10 million years before blowing itself to pieces. The shocking truth, which stares you in the face every time you look at a window, is that the whole Universe is founded on random chance. So upset was Einstein by this idea that he stuck out his lip and declared: "God does not play dice with the Universe!"

The trouble is He does. As British physicist Stephen Hawking has wryly pointed out: "Not only does God play dice with the Universe, he throws the dice where we cannot see them!"

When Einstein received the Nobel Prize for Physics in 1921 it was not for his more famous theory of relativity but for his explanation of the photoelectric effect. And this was no aberration on the part of the Nobel committee. Einstein himself considered his work on the "quantum" the only thing he ever did in science that was truly revolutionary. And the Nobel committee completely agreed with him.

Quantum theory, born out of the struggle to reconcile light and matter, was fundamentally at odds with all science that had gone before. Physics, pre-1900, was basically a recipe for predicting the future with absolute certainty. If a planet is in a particular place now, in a day's time it will have moved to another place, which can be predicted with 100 per cent confidence by using Newton's laws of motion and the law of gravity. Contrast this with an atom flying through space. Nothing is knowable with certainty. All we can ever predict is its probable path, its probable final position.

Whereas quantum is based on uncertainty, the rest of physics is based on certainty. To say this is a problem for physicists is a bit of an understatement! "Physics has given up on the problem of trying to predict what would happen in a given circumstance," said Richard Feynman. "We can only predict the odds."

All is not lost, however. If the microworld were totally unpredictable, it would be a realm of total chaos. But things are not this bad. Although what atoms and their like get up to is intrinsically unpredictable, it turns out that the unpredictability is at least predictable!

## PREDICTING THE UNPREDICTABILITY

Think of the window again. Each photon has a 95 per cent chance of being transmitted and a 5 per cent chance of being reflected. But what determines these probabilities?

Well, the two different pictures of light—as a particle and as a wave—must produce the same outcome. If half the wave goes through and half is reflected, the only way to reconcile the wave and particle pictures is if each individual particle of light has a 50 per cent probability of being transmitted and a 50 per cent probability of being reflected. Similarly, if 95 per cent of the wave is transmitted and 5 per cent is reflected, the corresponding probabilities for the transmission and reflection of individual photons must be 95 per cent and 5 per cent.

To get agreement between the two pictures of light, the particle-like aspect of light must somehow be "informed" about how to behave by its wavelike aspect. In other words, in the microscopic domain, waves do not simply behave like particles; those particles behave like waves as well! There is perfect symmetry. In fact, in a sense this statement is all you need to know about quantum theory (apart from a few details). Everything else follows unavoidably. All the weirdness, all the amazing richness of the microscopic world, is a direct consequence of this wave-particle "duality" of the basic building blocks of reality.

But how exactly does light's wavelike aspect inform its particle-like aspect about how to behave? This is not an easy question to answer.

Light reveals itself either as a stream of particles or as a wave. We never see both sides of the coin at the same time. So when we observe light as a stream of particles, there is no wave in existence to inform those particles about how to behave. Physicists therefore have a problem in explaining the fact that photons do things—for instance, fly through windows—as if directed by a wave.

They solve the problem in a peculiar way. In the absence of a real wave, they imagine an abstract wave—a mathematical wave. If this sounds ludicrous, this was pretty much the reaction of physicists when the idea was first proposed by the Austrian physicist Erwin Schrödinger in the 1920s. Schrödinger imagined an abstract mathematical wave that spread through space, encountering obstacles and being reflected and transmitted, just like a water wave spreading on a pond. In places where the height of the wave was large, the probability of finding a particle was highest, and in locations where it was small, the probability was lowest. In this way Schrödinger's wave of probability christened the wave function, informed a particle what to do, and not just a photon—any microscopic particle, from an atom to a constituent of an atom like an electron.

There is a subtlety here. Physicists could make Schrödinger's picture accord with reality only if the probability of finding a particle at any point was related to the square of the height of the probability wave at that point. In other words, if the probability wave at some point in space is twice as high as it is at another point in space, the particle is four times as likely to be found there than at the other place.

The fact that it is the square of the probability wave and not the probability wave itself that has real physical meaning to this day causes debate about whether the wave is a real thing lurking beneath the skin of the world or just a convenient mathematical device for calculating things. Most but not all people favour the latter.

The probability wave is crucially important because it makes a connection between the wavelike aspect of matter and familiar waves of all kinds, from water waves to sound waves to earthquake waves. All obey a so-called wave equation. This describes how they ripple through space and allows physicists to predict the wave height at any location at any time. Schrödinger's great triumph was to find the wave equation that described the behaviour of the probability wave of atoms and their like.

By using the Schrödinger equation, it is possible to determine the probability of finding a particle at any location in space at any time. For instance, it can be used to describe photons impinging on the obstacle of a windowpane and to predict the 95 per cent probability of finding one on the far side of the pane. In fact, the Schrödinger equation can be used to predict the probability of any particle, be it a photon or an atom, doing just about anything. It provides the crucial bridge to the microscopic world, allowing physicists to predict everything that happens there—if not with 100 per cent certainty, at least with predictable uncertainty!

Where is all this talk of probability waves leading? Well, the fact that waves behave like particles in the microscopic world leads unavoidably to the realisation that the microscopic world dances to an entirely different tune than that of the everyday world. It is governed by random unpredictability. This in itself was a shocking, confidence-draining blow to physicists and their belief in a predictable, clockwork universe. But this, it turns out, is only the beginning. Nature had many more shocks in store. The fact that waves not only behave as particles but also that those particles behave as waves leads to the realisation that all the things that familiar waves, like water waves and sound waves, can do, so too can the probability waves that inform the behaviour of atoms, photons, and their kin.

So what? Well, waves can do an awful lot of different things. And each of these things turns out to have a semi-miraculous consequence in the microscopic world. The most straightforward thing waves can do is exist as superpositions. Remarkably, this enables an atom to be in two places at once, the equivalent of you being in London and New York at the same time.

# 3

# THE SCHIZOPHRENIC ATOM

### HOW AN ATOM CAN BE IN MANY PLACES AT ONCE AND DO MANY THINGS AT ONCE

*If you imagine the difference between an abacus and the world's fastest supercomputer, you would still not have the barest inkling of how much more powerful a quantum computer could be compared with the computers we have today.*

Julian Brown

*It's 2041. A boy sits at a computer in his bedroom. It's not an ordinary computer. It's a quantum computer. The boy gives the computer a task . . . and instantly it splits into thousands upon thousands of versions of itself, each of which works on a separate strand of the problem. Finally, after just a few seconds, the strands come back together and a single answer flashes on the computer display. It's an answer that all the normal computers in the world put together would have taken a trillion trillion years to find. Satisfied, the boy shuts the computer down and goes out to play, his night's homework done.*

Surely, no computer could possibly do what the boy's computer has just done? Not only could a computer do such a thing, crude versions are already in existence today. The only thing in serious dispute is whether such a quantum computer merely behaves like a vast multiplicity of computers or whether, as some believe, it literally exploits the computing power of multiple copies of itself existing in parallel realities, or universes.

The key property of a quantum computer—the ability to do many calculations at once—follows directly from two things that waves—and therefore microscopic particles such as atoms and photons, which behave like waves—can do. The first of those things can be seen in the case of ocean waves.

On the ocean there are both big waves and small ripples. But as anyone who has watched a heavy sea on a breezy day knows, you can also get big, rolling waves with tiny ripples superimposed on them. This is a general property of all waves. If two different waves can exist, so too can a combination, or superposition, of the waves. The fact that superpositions can exist is pretty innocuous in the everyday world. However, in the world of atoms and their constituents, its implications are nothing short of earth-shattering.

Think again of a photon impinging on a windowpane. The photon is informed about what to do by a probability wave, described by the Schrödinger equation. Since the photon can either be transmitted or reflected, the Schrödinger equation must permit the existence of two waves—one corresponding to the photon going through the window and another corresponding to the photon bouncing back. Nothing surprising here. However, remember that, if two waves are permitted to exist, a superposition of them is also permitted to exist. For waves at sea such a combination is nothing out of the ordinary. But here the combination corresponds to something quite extraordinary—the photon being both transmitted and reflected. In other words, the photon can be on both sides of the windowpane simultaneously!

And this unbelievable property follows unavoidably from just two facts: that photons are described by waves and that superpositions of waves are possible.

This is no theoretical fantasy. In experiments it is actually possible to observe a photon or an atom being in two places at once—the everyday equivalent of you being in San Francisco and Sydney at the same time. (More accurately, it is possible to observe the *consequences* of a photon or an atom being in two places at once.) And since there is no limit to the number of waves that can be superposed, a photon or an atom can be in three places, 10 places, a million places at once.

But the probability wave associated with a microscopic particle does more than inform it where it could be *located*. It informs it *how to behave* in all circumstances—telling a photon, for instance, whether or not to be transmitted or reflected by a pane of glass. Consequently, atoms and their like can not only be in many places at once, they can *do many things at once*, the equivalent of you cleaning the house, walking the dog, and doing the weekly supermarket shopping all at the same time. This is the secret behind the prodigious power of a quantum computer. It exploits the ability of atoms to do many things at once, to do many calculations at once.

## DOING MANY THINGS AT ONCE

The basic elements of a conventional computer are transistors. These have two distinct voltage states, one of which is used to represent the binary digit, or bit, "0", the other to represent a "1." A row of such zeros and ones can represent a large number, which in the computer can be added, subtracted, multiplied, and divided by another large number.[1] But in a quantum computer the basic elements—which may be single atoms—can be in a superposition of states. In other words, they can represent a zero and a one simultaneously. To distinguish them from normal bits, physicists call such schizophrenic entities quantum bits, or qubits.

---

[1] Binary was invented by the 17th-century mathematician Gottfried Leibniz. It is a way of representing numbers as a strings of zeros and ones. Usually, we use decimal, or base 10. The right-hand digit represents the ones, the next digit the tens, the next the $10 \times 10$s, and so on. So, for instance, 9,217 means $7 + 1 \times 10 + 2 \times (10 \times 10) + 9 \times (10 \times 10 \times 10)$. In binary, or base 2, the right-hand digit represents the ones, the next digit the twos, the next the $2 \times 2$s, and so on. So for instance, 1101 means $1 + 0 \times 2 + 1 \times (2 \times 2) + 1 \times (2 \times 2 \times 2)$, which in decimal is 13.

One qubit can be in two states (0 or 1), two qubits in four (00 or 01 or 10 or 11), three qubits in eight, and so on. Consequently, when you calculate with a single qubit, you can do two calculations simultaneously, with two qubits four calculations, with three eight, and so on. If this doesn't impress you, with 10 qubits you could do 1,024 calculations all at once, with 100 qubits 100 billion billion billion! Not surprisingly, physicists positively salivate at the prospect of quantum computers. For some calculations, they could massively outperform conventional computers, making conventional personal computers appear positively retarded.

But for a quantum computer to work, wave superpositions are not sufficient on their own. They need another essential wave ingredient: interference.

The interference of light observed by Thomas Young in the 18th century was the key observation that convinced everyone that light was a wave. When, at the beginning of the 20th century, light was also shown to behave like a stream of particles, Young's double slit experiment assumed a new and unexpected importance—as a means of exposing the central peculiarity of the microscopic world.

## INTERFERENCE IS THE KEY

In the modern incarnation of Young's experiment, a double slit in an opaque screen is illuminated with light, which is undeniably a stream of particles. In practice, this means using a light source so feeble that it spits out photons one at a time. Sensitive detectors at different positions on the second screen count the arrival of photons. After the experiment has been running for a while, the detectors show something remarkable. Some places on the screen get peppered with photons while other places are completely avoided. What is more, the places that are peppered by photons and the places that are avoided alternate, forming vertical stripes—exactly as in Young's original experiment.

But wait a minute! In Young's experiment the dark and light

bands are caused by interference. And a fundamental feature of interference is that it involves the mingling of two sets of waves from the same source—the light from one slit with the light from the other slit. But in this case the photons are arriving at the double slit one at a time. Each photon is completely alone, with no other photon to mingle with. How, then, can there be any interference? How can it know where its fellow photons will land?

There would appear to be only one way—if each photon somehow goes through both slits simultaneously. Then it can interfere with itself. In other words, each photon must be in a superposition of two states—one a wave corresponding to a photon going through the left-hand slit and the other a wave corresponding to a photon going through the right-hand slit.

The double slit experiment can be done with photons or atoms or any other microscopic particles. It shows graphically how the behaviour of such particles—where they can and cannot strike the second screen—is orchestrated by their wavelike alter ego. But this is not all the double slit experiment demonstrates. Crucially, it shows that the individual waves that make up a superposition are not passive but can actively interfere with each other. It is this ability of the individual states of a superposition to interfere with each other that is the absolute key to the microscopic world, spawning all manner of weird quantum phenomena.

Take quantum computers. The reason they can carry out many calculations at once is because they can exist in a superposition of states. For instance, a 10-element quantum computer is simultaneously in 1,024 states and can therefore carry out 1,024 calculations at once. But all the parallel strands of a calculation are of absolutely no use unless they get woven together. Interference is the means by which this is accomplished. It is the means by which the 1,024 states of the superposition can interact and influence each other. Because of interference, the single answer coughed out by the quantum computer is able to reflect and synthesise what was going on in all those 1,024 parallel calculations.

Think of a problem divided into 1,024 separate pieces and one person working on each piece. For the problem to be solved, the 1,024 people must communicate with each other and exchange results. This is what interference makes possible in a quantum computer.

An important point worth making here is that, although superpositions are a fundamental feature of the microscopic world, it is a curious property of reality that they are never actually observed. All we ever see are the consequences of their existence—what results when the individual waves of a superposition *interfere* with each other. In the case of the double slit experiment, for instance, all we ever see is an interference pattern, from which we infer that an electron was in a superposition in which it went through both slits simultaneously. It is impossible to actually *catch* an electron going through both slits at once. This is what was meant by the earlier statement that it is possible only to observe the *consequences* of an atom being in two places at once, not it actually being in two places at once.

## MULTIPLE UNIVERSES

The extraordinary ability of quantum computers to do enormous numbers of calculations simultaneously poses a puzzle. Though practical quantum computers are currently at a primitive stage, manipulating only a handful of qubits, it is nevertheless possible to imagine a quantum computer that can do billions, trillions, or quadrillions of calculations simultaneously. In fact, it is quite possible that in 30 or 40 years we will be able to build a quantum computer that can do more calculations simultaneously than there are particles in the Universe. This hypothetical situation poses a sticky question: Where exactly will such a computer be doing its calculations? After all, if such a computer can do more calculations simultaneously than there are particles in the Universe, it stands to reason that the Universe has insufficient computing resources to carry them out.

One extraordinary possibility, which provides a way out of the conundrum, is that a quantum computer does its calculations in

parallel realities or universes. The idea goes back to a Princeton gradu-
ate student named Hugh Everett III, who, in 1957, wondered why
quantum theory is such a brilliant description of the microscopic
world of atoms but we never actually see superpositions. Everett's
extraordinary answer was that each state of the superposition exists
in a totally separate reality. In other words, there exists a multiplicity
of realities—a *multiverse*—where all possible quantum events occur.

Although Everett proposed his "Many Worlds" idea long before
the advent of quantum computers, it can shed some helpful light on
them. According to the Many Worlds idea, when a quantum computer is given a problem, it splits into multiple versions of itself, each
living in a separate reality. This is why the boy's quantum personal
computer at the start of this chapter split into so many copies. Each
version of the computer works on a strand of the problem, and the
strands are brought together by interference. In Everett's picture,
therefore, interference has a very special significance. It is the all-
important *bridge* between separate universes, the means by which
they interact and influence each other.

Everett had no idea *where* all the parallel universes were located.
And, frankly, nor do the modern-day proponents of the Many Worlds
idea. As Douglas Adams wryly observed in *The Hitchhiker's Guide to
the Galaxy:* "There are two things you should remember when deal-
ing with parallel universes. One, they're not really parallel, and two,
they're not really universes!"

Despite such puzzles, half a century after Everett proposed the
Many Worlds idea, it is undergoing an upsurge in popularity. An increasing number of physicists, most notably David Deutsch of the
University of Oxford, are taking it seriously. "The quantum theory of
parallel universes is not some troublesome, optional interpretation
emerging from arcane theoretical considerations," says Deutsch in his
book, *The Fabric of Reality.* "It is *the* explanation—the only one that is
tenable—of a remarkable and counterintuitive reality."

If you go along with Deutsch—and the Many Worlds idea predicts exactly the same outcome for every conceivable experiment as

more conventional interpretations of quantum theory—then quantum computers are something radically new under the Sun. They are the very first machines humans have ever built that exploit the resources of multiple realities. Even if you do not believe the Many Worlds idea, it still provides a simple and intuitive way of imagining what is going on in the mysterious quantum world. For instance, in the double slit experiment, it is not necessary to imagine a single photon going through both slits simultaneously and interfering with itself. Instead, a photon going through one slit interferes with another photon going through the other slit. What other photon, you may ask? A photon in a neighbouring universe, of course!

## WHY ARE ONLY SMALL THINGS QUANTUM?

Quantum computers are extremely difficult to build. The reason is that the ability of the individual states of a quantum superposition to interfere with each other is destroyed, or severely degraded, by the environment. This destruction can be vividly seen in the double slit experiment.

If some kind of particle detector is used to spot a particle going through one of the slits, the interference stripes on the screen immediately vanish, to be replaced by more or less uniform illumination. The act of observing which slit the particle goes through is all that is needed to destroy the superposition in which it goes through both slits simultaneously. And a particle going through one slit only is as likely to exhibit interference as you are to hear the sound of one hand clapping.

What has really happened here is that an attempt has been made to locate, or measure, the particle by the outside world. Knowledge of the superposition by the outside world is all that is needed to destroy it. It is almost as if quantum superpositions are a secret. Of course, once the world knows about the secret, the secret no longer exists!

Superpositions are *continually* being measured by their environment. And it takes only a single photon to bounce off a superposition and take information about it to the rest of the world to destroy the superposition. This process of natural measurement is called decoherence. It is the ultimate reason we do not see weird quantum behaviour in the everyday world.[2] Although naively we may think of quantum behaviour as a property of small things like atoms but not of big things like people and trees, this is not necessarily so. Quantum behaviour is actually a property of isolated things. The reason we see it in the microscopic world but not in the everyday world is simply because it is easier to isolate a small thing from its surroundings than a big thing.

The price of quantum schizophrenia is therefore isolation. As long as a microscopic particle like an atom can remain isolated from the outside world, it can do many different things at once. This is not difficult in the microscopic world, where quantum schizophrenia is an everyday phenomenon. However, in the large-scale world in which we live, it is nearly impossible, with countless quadrillions of photons bouncing off every object every second.

Keeping a quantum computer isolated from its surroundings is the main obstacle facing physicists in trying to construct such a machine. So far, the biggest quantum computer they have managed to build has been composed of only 10 atoms, storing 10 qubits. Keeping 10 atoms isolated from their surroundings for any length of time takes all their ingenuity. If a single photon bounces off the computer, 10 schizophrenic atoms instantly become 10 ordinary atoms.

---

[2] I am totally aware that all this talk of quantumness being a "secret" that is destroyed if the rest of the world learns about it is a complete fudge. But it is sufficient for our discussion here. Decoherence, the means by which the quantum world, with its schizophrenic superpositions, becomes the everyday world where trees and people are never in two places at once, is a can of worms with which the experts are still wrestling. For a real explanation, see Chapter 5, "The Telepathic Universe."

Decoherence illustrates a limitation of quantum computers not often publicised amid the hype surrounding such devices. To extract an answer, someone from the outside world—you—must interact with it, and this necessarily destroys the superposition. The quantum computer reverts to being an ordinary computer in a single state. A 10-qubit machine, instead of spitting out the answers to 1,024 separate calculations, spits out just one.

Quantum computers are therefore restricted to parallel calculations that output only a single answer. Consequently, only a limited number of problems are suited to solution by quantum computer, and much ingenuity is required to find them. They are not, as is often claimed, the greatest thing since sliced bread. Nevertheless, when a problem is found that plays to the strengths of a quantum computer, it can massively outperform a conventional computer, calculating in seconds what otherwise might take longer than the lifetime of the Universe.

On the other hand, decoherence, which is the greatest enemy of those struggling to build quantum computers, is also their greatest friend. It is because of decoherence, after all, that the giant superposition of a quantum computer with all its mutually interfering strands is finally destroyed; it is only by being destroyed—reduced to a single state representing a single answer—that anything useful comes out of such a machine. The world of the quantum is indeed a paradoxical one!

# 4

# UNCERTAINTY AND THE LIMITS OF KNOWLEDGE

WHY WE CAN NEVER KNOW ALL WE WOULD LIKE TO KNOW ABOUT ATOMS
AND WHY THIS FACT MAKES ATOMS POSSIBLE

*Passing farther through the quantum land our travelers met quite a lot
of other interesting phenomena, such as quantum mosquitoes, which
could scarcely be located at all, owing to their small mass.*

George Gamow

*He must be going mad. Only moments before he had parked his shiny
red Ferrari in the garage. He had even stood there on the driveway, ad-
miring his pride and joy until the last possible moment, as the automatic door swung shut. But then as he crunched across the gravel to his
front door there had been a curious rustling of the air, a faint tremor of
the ground. He had wheeled round. And there, squatting back on his
driveway, in front of the still-locked garage doors, was his beautiful red
Ferrari!*

Such Houdini-like feats of escapology are never of course seen in the
everyday world. In the realm of the ultrasmall, however, they are a
common occurrence. One instant an atom can be locked up in a microscopic prison; the next it has shed its shackles and slipped away
silently into the night.

This miraculous ability to escape escape-proof prisons is entirely due to the wavelike face of microscopic particles, which enables atoms and their constituents to do all the things that waves can do. And one of the many things waves can do is penetrate apparently impenetrable barriers. This is not an obvious or well-known wave property. But it can be demonstrated by a light beam travelling through a block of glass and trying to escape into the air beyond.

The key thing is what happens at the edge of the glass block, the boundary where the glass meets the air. If the light happens to strike the boundary at a shallow angle, it gets reflected back into the glass block and fails to escape into the air beyond. In effect, it is imprisoned in the glass. However, something radically different happens if another block of glass is brought close to the boundary, leaving a small gap of air between the two blocks. Just as before, some of the light is reflected back into the glass. But—and this is the crucial thing—some of the light now leaps the air gap and travels into the second glass block.

The parallel between the Ferrari escaping its garage and the light escaping the block of glass may not be very obvious. However, for all intents and purposes, the air gap should be just as impenetrable a barrier to the light as the garage walls are to the Ferrari.

The reason the light wave can penetrate the barrier and escape from the block of glass is that a wave is not a localised thing but something spread out through space. So when the light waves strike the glass-air boundary and are reflected back into the glass, they are not actually reflected from the exact boundary of the glass. Instead, they penetrate a short distance into the air beyond. Consequently, if they encounter another block of glass before they can turn back, they can continue on their way. Place a second glass block within a hair's breadth of the first and, hey presto, the light jumps the air gap and escapes its prison.

This ability to penetrate an apparently impenetrable barrier is common to all types of waves, from light waves to sound waves to the probability waves associated with atoms. It therefore manifests itself in the microscopic world. Arguably, the most striking example is the phenomenon of alpha decay in which an alpha particle breaks out of the apparently escape-proof prison of an atomic nucleus.

## BREAKING OUT OF A NUCLEUS

An alpha particle is the nucleus of a helium atom. An unstable, or radioactive, nucleus sometimes spits out an alpha particle in a desperate attempt to turn itself into a lighter and more stable nucleus. The process poses a big puzzle, however. By rights, an alpha particle should not be able to get out of a nucleus.

Think of an Olympic high jumper penned in by a 5-metre-high metal fence. Even though he is one of the best high jumpers in the world, there is no way he can jump over a fence that high. No human being alive has sufficient strength in their legs. Well, an alpha particle inside an atomic nucleus finds itself in a similar position. The barrier that pens it in is created by the nuclear forces that operate inside a nucleus, but it is just as impenetrable a barrier to the alpha particle as the solid metal fence is to the high jumper.

Contrary to all expectations, however, alpha particles do escape from atomic nuclei. And their escape is entirely due to their wavelike face. Like light waves trapped in a glass block, they can penetrate an apparently impenetrable barrier and slip away quietly into the outside world.

This process is called quantum tunnelling and alpha particles are said to "tunnel" out of an atomic nucleus. Tunnelling is actually an instance of a more general phenomenon known as uncertainty, which puts a fundamental limit on what we can and cannot know about the microscopic world. The double slit experiment is an excellent demonstration of uncertainty.

## THE HEISENBERG UNCERTAINTY PRINCIPLE

The reason a microscopic particle like an electron can go through both slits in the screen simultaneously is that it can exist as a superposition of two waves—one wave corresponding to the particle going through one slit and the other to the particle going through the other slit. But that is not sufficient to guarantee that its schizophrenic behaviour will be noticed. For that to happen, an interference pattern must appear on the second screen. But this, of course, requires the individual waves in the superposition to interfere. The fact that interference is a crucial ingredient for the electron to exhibit weird quantum behaviour turns out to have profound implications for what nature permits us to know about the electron.

Say in the double slit experiment we try to locate the slit each electron goes through. If we succeed, the interference pattern on the second screen disappears. After all, interference requires that two things mingle. If the electron and its associated probability wave go through only one slit, there is only one thing.

How, in practice, could we locate which slit an electron goes through? Well, to make the double slit experiment a bit easier to visualise, think of an electron as a bullet from a machine gun and the screen as a thick metal sheet with two vertical parallel slits. When bullets are fired at the screen, some enter the slits and go through. Think of the slits as deep channels cut through the thick metal. The bullets ricochet off the internal walls of the channels and by this means reach the second screen. They can obviously hit any point on the second screen. But, for simplicity, imagine they end up at the midpoint of the second screen. Also for simplicity, say that at this point the probability waves associated with the bullets interfere constructively, so it is a place that gets peppered with lots of bullets.

Now, when a bullet ricochets off the inside of a slit, it causes the metal screen to recoil in the opposite direction. It's the same if you are playing tennis and a fast serve ricochets off your racquet. Your racquet recoils in the opposite direction. Crucially, the recoil of the screen can be used to deduce which slit a bullet goes through. After all, if the screen moves to the left, the bullet must have gone through the left-hand slit; if it moves to the right, it must have been the right-hand slit.

However, we know that if we locate which slit a bullet goes through, it destroys the interference pattern on the second screen. This is straightforward to understand from the wave point of view. We are as unlikely to see one thing interfere with itself as we are to hear the sound of one hand clapping. But how do we make sense of things from the equally valid particle point of view?

Remember that the interference pattern on the second screen is like a supermarket bar code. It consists of vertical "stripes" where no bullets hit, alternating with vertical stripes where lots of bullets hit. For simplicity, think of the stripes as black and white. The key question therefore is: From the bullet's point of view, what would it take to destroy the interference pattern?

The answer is a little bit of sideways jitter. If each bullet, instead of flying unerringly towards a black stripe, possesses a little sideways jitter in its trajectory so that it can hit either the black stripe or an adjacent white stripe, this will be sufficient to "smear out" the interference pattern. Stripes that were formerly white will become blacker, and stripes that were formerly black will become whiter. The net result will be a uniform gray. The interference pattern will be smeared out.

Because it must be impossible to tell whether a given bullet will hit a black stripe or an adjacent white stripe (or vice versa), the jittery sideways motion of each bullet must be entirely unpredictable. And all this must come to pass for no other reason than that we are locating which slit each bullet goes through by the recoil of the screen.

In other words, the very act of pinning down the location of a particle like an electron adds unpredictable jitter, making its velocity uncertain. And the opposite is true as well. The act of pinning down the velocity of a particle makes its location uncertain. The first person to recognise and quantify this effect was the German physicist Werner Heisenberg, and it is called the Heisenberg uncertainty principle in his honour.

According to the uncertainty principle, it is impossible to know both the location and the velocity of a microscopic particle with complete certainty. There is a trade-off, however. The more precisely its location is pinned down, the more uncertain is its velocity. And the more precisely its velocity is pinned down, the more uncertain its location.

Imagine if this constraint also applied to what we could know about the everyday world. If we had precise knowledge of the speed of a jet aeroplane, we would not be able to tell whether it was over London or New York. And if we had precise knowledge of the location of the aeroplane, we would be unable to tell whether it was cruising at 1,000 kilometres per hour or 1 kilometre per hour—and about to plummet out of the sky.

The uncertainty principle exists to *protect* quantum theory. If you could measure the properties of atoms and their like better than the uncertainty principle permits, you would destroy their wave behaviour—specifically, interference. And without interference, quantum theory would be impossible. Measuring the position and velocity of a particle with greater accuracy than the uncertainty principle dictates must therefore be impossible. Because of the Heisenberg uncertainty principle, when we try to look closely at the microscopic world, it starts to get fuzzy, like a newspaper picture that has been overmagnified. Infuriatingly, nature does not permit us to measure precisely all we would like to measure. There is a limit to our knowledge.

This limit is not simply a quirk of the double slit experiment. It is fundamental. As Richard Feynman remarked: "No one has ever found (or even thought of) a way around the uncertainty principle. Nor are they ever likely to."

It is because alpha particles have a wavelike character that they can escape the apparently escape-proof prison of an atomic nucleus.

However, the Heisenberg uncertainty principle makes it possible to understand the phenomenon from the particle point of view.

## GOING WHERE NO HIGH JUMPER HAS GONE BEFORE

Recall that an alpha particle in a nucleus is like an Olympic high jumper corralled by a 5-metre-high fence. Common sense says that it is moving about inside the nucleus with insufficient speed to launch itself over the barrier. But common sense applies only to the everyday world, not to the microscopic world. Ensnared in its nuclear prison, the alpha particle is very localised in space—that is, its position is pinned down with great accuracy. According to the Heisenberg uncertainty principle, then, its velocity must necessarily be very uncertain. It could, in other words, be much greater than we think. And if it is greater, then, contrary to all expectations, the alpha particle can leap out of the nucleus—a feat comparable to the Olympic high jumper jumping the 5-metre fence.

Alpha particles emerge into the world outside their prison as surprisingly as the Ferrari emerged into the world outside its garage. And this "tunnelling" is due to the Heisenberg uncertainty principle. But tunnelling is a two-way process. Not only can subatomic particles like alpha particles tunnel out of a nucleus, they can tunnel into it too. In fact, such tunnelling in reverse helps explain a great mystery: why the Sun shines.

## TUNNELLING IN THE SUN

The Sun generates heat by gluing together protons—the nuclei of hydrogen atoms—to make the nuclei of helium atoms.[1] This nuclear fusion produces as a by-product a dam burst of nuclear binding energy, which ultimately emerges from the Sun as sunlight.

---

[1]See Chapter 8, "$E = mc^2$ and the Weight of Sunshine."

But hydrogen fusion has a problem. The force of attraction that glues together protons—the "strong nuclear force"—has an extremely short range. For two protons in the Sun to come under its influence and be snapped together, they must pass extremely close to each other. But two protons, by virtue of their similar electric charge, repel each other ferociously. To overcome this fierce repulsion, the protons must collide at enormous speed. In practice, this requires the core of the Sun, where nuclear fusion goes on, to be at an extremely high temperature.

Physicists calculated the necessary temperature in the 1920s, just as soon as it was suspected that the Sun was running on hydrogen fusion. It turned out to be roughly 10 billion degrees. This, however, posed a problem. The temperature at the heart of the Sun was known to be only about 15 million degrees—roughly a thousand times lower. By rights, the Sun should not be shining at all. Enter the German physicist Fritz Houtermans and the English astronomer Robert Atkinson.

When a proton in the core of the Sun approaches another proton and is pushed back by its fierce repulsion, it is just as if it encounters a high brick wall surrounding the second proton. At the 15 million degrees temperature in the heart of the Sun, the proton would appear to be moving far too slowly to jump the wall. However, the Heisenberg uncertainty principle changes everything.

In 1929, Houtermans and Atkinson carried out the relevant calculations. They discovered that the first proton can tunnel through the apparently impenetrable barrier around the second proton and successfully fuse with it even at the ultralow temperature of 15 million degrees. What is more, this explains perfectly the observed heat output of the Sun.

The night after Houtermans and Atkinson did the calculation, Houtermans reportedly tried to impress his girlfriend with a line that nobody in history had used before. As they stood beneath a perfect moonless sky, he boasted that he was the only person in the world who knew why the stars were shining. It must have worked. Two years later, Charlotte Riefenstahl agreed to marry him. (Actually, she married him twice, but that's another story.)

Sunlight apart, the Heisenberg uncertainty principle explains something much closer to home: the very existence of the atoms in our bodies.

## UNCERTAINTY AND THE EXISTENCE OF ATOMS

By 1911 the Cambridge experiments of New Zealand physicist Ernest Rutherford had revealed the atom as resembling a miniature solar system. Tiny electrons flitted about a compact atomic nucleus much like planets around the Sun. However, according to Maxwell's theory of electromagnetism, an orbiting electron should radiate light energy and, within a mere hundred-millionth of a second, spiral into the nucleus. "Atoms," as Richard Feynman pointed out, "are completely impossible from the classical point of view." But atoms do exist. And the explanation comes from quantum theory.

An electron cannot get too close to a nucleus because, if it did, its location in space would be very precisely known. But according to the Heisenberg uncertainty principle, this would mean that its velocity would be very uncertain. It could become enormously huge.

Imagine an angry bee in a shrinking box. The smaller the box gets, the angrier the bee and the more violently it batters itself against the walls of its prison. This is pretty much the way an electron behaves in an atom. If it were squeezed into the nucleus itself, it would acquire an enormous speed—far too great to stay confined in the nucleus.

The Heisenberg uncertainty principle, which explains why electrons do not spiral into their nuclei, is therefore the ultimate reason why the ground beneath our feet is solid. But the principle does more than simply explain the existence of atoms and the solidity of matter. It explains why atoms are so big—or at least so much bigger than the nuclei at their cores.

## WHY ATOMS ARE SO BIG

Recall that a typical atom is about 100,000 times bigger than the nucleus at its centre. Understanding why there is such a fantastic amount of empty space in atoms requires being a bit more precise about the Heisenberg uncertainty principle. Strictly speaking, it says that it is a particle's position and momentum—rather than just its velocity—that cannot simultaneously be determined with 100 per cent certainty.

The momentum of a particle is the product of its mass and velocity. It's really just a measure of how difficult it is to stop something that is moving. A train, for instance, has a lot of momentum compared to a car, even if the car is going faster. A proton in an atomic nucleus is about 2,000 times more massive than an electron. According to the Heisenberg uncertainty principle, then, if a proton and an electron are confined in the same volume of space, the electron will be moving about 2,000 times faster.

Already, we get an inkling of why the electrons in an atom must have a far bigger volume to fly about in than the protons and neutrons in the nucleus. But atoms are not just 2,000 times bigger than their nuclei; they are more like 100,000 times bigger. Why?

The answer is that an electron in an atom and a proton in a nucleus are not in the grip of the same force. While the nuclear particles are held by the powerful "strong nuclear" force, the electrons are held by the much weaker electric force. Think of the electrons flying about the nucleus attached to gossamer threads of elastic while the protons and the neutrons are constrained by elastic 50 times thicker. Here is the explanation for why the atom is a whopping 100,000 times bigger than the nucleus.

But the electrons in an atom do not orbit at one particular distance from the nucleus. They are permitted to orbit at a range of distances. Explaining this requires resorting to yet another wave picture—this one involving organ pipes!

## OF ATOMS AND ORGAN PIPES

There are always many different ways of looking at things in the quantum world, each a glimpse of a truth that is frustratingly elusive. One way is to think of the probability waves associated with an atom's electrons as being like sound waves confined to an organ pipe. It is not possible to make just any note with the organ pipe. The sound can vibrate in only a limited number of different ways, each with a definite pitch, or frequency.

This turns out to be a general property of waves, not just sound waves. In a confined space they can exist only at particular, definite frequencies.

Now think of an electron in an atom. It behaves like a wave. And it is gripped tightly by the electrical force of the atomic nucleus. This may not be exactly the same as being trapped in a physical container. However, it confines the electron wave as surely as the wall of an organ pipe confines a sound wave. The electron wave can therefore exist at only certain frequencies.

The frequencies of the sound waves in an organ pipe and of the electron waves in an atom depend on the characteristics of the organ pipe—a small organ pipe, for instance, produces higher-pitched notes than a big organ pipe—and on the characteristics of the electrical force of the atomic nucleus. In general, though, there is lowest, or fundamental, frequency and a series of higher-frequency "overtones."

A higher-frequency wave has more peaks and troughs in a given space. It is choppier, more violent. In the case of an atom, such a wave corresponds to a faster-moving, more energetic electron. And a faster-moving, more energetic electron is able to defy the electrical attraction of the nucleus and orbit farther away.

The picture that emerges is of an electron that is permitted to orbit at only certain special distances from the nucleus. This is quite unlike our solar system where a planet such as Earth could, in principle, orbit at any distance whatsoever from the Sun.

This property highlights another important difference between the microscopic world of atoms and the everyday world. In the everyday world, all things are continuous—a planet can orbit the Sun anywhere it likes, people can be any weight they like—whereas things in the microscopic world are discontinuous—an electron can exist in only certain orbits around a nucleus, light and matter can come in only certain indivisible chunks. Physicists call the chunks quanta—which is why the physics of the microscopic world is known as quantum theory.

The innermost orbit of an electron in an atom is determined by the Heisenberg uncertainty principle—by its hornetlike resistance to being confined in a small space. But the Heisenberg uncertainty principle does not simply prevent small things like atoms from shrinking without limit—ultimately explaining the solidity of matter. It also prevents far bigger things from shrinking without limit. The far bigger things in question are stars.

## UNCERTAINTY AND STARS

A star is a giant ball of gas held together by the gravitational pull of its own matter. That pull is constantly trying to shrink the star and, if unopposed, would very quickly collapse it down to the merest speck—a black hole. For the Sun this would take less than half an hour. Since the Sun is very definitely not shrinking down to a speck, there must be another force counteracting gravity. There is. It comes from the hot matter inside. The Sun—along with every other normal star—is in a delicate state of balance, with the inward force of gravity exactly matched by the outward force of its hot interior.

This balance, however, is temporary. The outward force can be maintained only while there is fuel to burn and keep the star hot. Sooner or later, the fuel will run out. For the Sun this will occur in about another 5 billion years. When this happens, gravity will be king. Unopposed, it will crush the star, shrinking it ever smaller.

But all is not lost. In the dense, hot environment inside a star, frequent and violent collisions between high-speed atoms strip them of their electrons, creating a plasma, a gas of atomic nuclei mixed in with a gas of electrons. It is the tiny electrons that unexpectedly come to the rescue of the fast-shrinking star. As the electrons in the star's matter are jammed ever closer together, they buzz about ever more violently because of the Heisenberg uncertainty principle. They batter anything trying to confine them, and this collective battering results in a tremendous outward force. Eventually, it is enough to slow and halt the shrinkage of the star.

A new balance is struck with the inward pull of gravity balanced not by the outward force of the star's hot matter but by the naked force of its electrons. Physicists call it degeneracy pressure. But it's just a fancy term for the resistance of electrons to being squeezed too close together. A star supported against gravity by electron pressure is known as a white dwarf. Little more than the size of Earth and occupying about a millionth of the star's former volume, a white dwarf is an enormously dense object. A sugarcube of its matter weighs as much as a car!

One day the Sun will become a white dwarf. Such stars have no means of replenishing their lost heat. They are nothing more than stellar embers, cooling inexorably and gradually fading from view. But the electron pressure that prevents white dwarfs from shrinking under their own gravity has its limits. The more massive a star, the stronger its self-gravity. If the star is massive enough, its gravity will be powerful enough to overcome even the stiff resistance of the star's electrons.

In fact, the star is sabotaged from both outside and inside. The stronger the gravity of a star, the more it squeezes the gas inside. And the more a gas is squeezed, the hotter it gets, as anyone who has used a bicycle pump knows. Since heat is nothing more than the microscopic jiggling of matter, the electrons inside the star fly about ever faster—so fast, in fact, that the effects of relativity become important.[2] The electrons get more massive rather than much faster, which means they are less effective at battering the walls of their prison.

The star suffers a double whammy—crushed by stronger gravity and simultaneously robbed of the ability to fight back. The two effects combine to ensure that the heaviest a white dwarf can be is a mere 40 per cent more massive than the Sun. If a star is heavier than this "Chandrasekhar limit", electron pressure is powerless to halt its headlong collapse and it just goes on shrinking.

But, once again, all is not lost. Eventually, the star shrinks so much that its electrons, despite their tremendous aversion to being confined in a small volume, are actually squeezed into the atomic nuclei. There they react with protons to form neutrons, so that the whole star becomes one giant mass of neutrons.

Recall that all particles of matter—not just electrons—resist being confined because of the Heisenberg uncertainty principle. Neutrons are thousands of times more massive than electrons. They therefore have to be squeezed into a volume thousands of times smaller to begin to put up significant resistance. In fact, they have to be squeezed together until they are virtually touching before they finally halt the shrinkage of the star.

A star supported against gravity by neutron degeneracy pressure is known as a neutron star. In effect, it is a huge atomic nucleus with all the empty space squeezed out of its matter. Since atoms are mostly empty space, with their nuclei 100,000 times smaller than their surrounding cloud of orbiting electrons, neutron stars are 100,000 times smaller than a normal star. This makes them only about 15 kilometres across, not much bigger than Mount Everest. So dense is a neutron star that a sugarcube of its matter weighs as much as the entire human race. (This, of course, is an illustration of just how much empty space there is in all of us. Squeeze it all out and humanity would fit in your hand.)

_____

[2]See Chapter 7, "The Death of Space and Time."

Such stars are thought to form violently in supernova explosions. While the outer regions of a star are blown into space, the inner core shrinks to form a neutron star. Neutron stars, being tiny and cold, ought to be difficult to spot. However, they are born spinning very fast and produce lighthouse beams of radio waves that flash around the sky. Such pulsating neutron stars, or simply pulsars, semaphore their existence to astronomers.

## UNCERTAINTY AND THE VACUUM

White dwarfs and neutron stars apart, perhaps the most remarkable consequence of the Heisenberg uncertainty principle is the modern picture of empty space. It simply cannot be empty!

The Heisenberg uncertainty principle can be reformulated to say that it is impossible to simultaneously measure the energy of a particle and the interval of time for which it has been in existence. Consequently, if we consider what happens in a region of empty space in a very tiny interval of time, there will be a large uncertainty in the energy content of that region. In other words, energy can appear out of nothing!

Now, mass is a form of energy.[3] This means that mass too can appear out of nothing. The proviso is that it can appear only for a mere split second before disappearing again. The laws of nature, which usually prevent things from appearing out of nothing, appear to turn a blind eye to events that happen too quickly. It's rather like a teenager's dad not noticing his son has borrowed the car for the night as long as it gets put back in the garage before daybreak.

In practice, mass is conjured out of empty space in the form of microscopic particles of matter. The quantum vacuum is actually a seething morass of microscopic particles such as electrons popping

---

[3]See Chapter 8, "$E = mc^2$ and the Weight of Sunlight."

into existence and then vanishing again.[4] And this is no mere theory. It actually has observable consequences. The roiling sea of the quantum vacuum actually buffets the outer electrons in atoms, very slightly changing the energy of the light they give out.[5]

The fact that the laws of nature permit something to come out of nothing has not escaped cosmologists, people who think about the origin of the Universe. Could it be, they wonder, that the entire Universe is nothing more than a quantum fluctuation of the vacuum? It's an extraordinary thought.

---

[4]Actually, every particle created is created alongside its antiparticle, a particle with opposite properties. So a negatively charged electron is always created with a positively charged positron.

[5]This effect is called the Lamb shift.

# 5

# THE TELEPATHIC UNIVERSE

HOW ATOMS CAN INFLUENCE EACH OTHER INSTANTLY
EVEN WHEN ON OPPOSITE SIDES OF THE UNIVERSE

*Beam me up, Mr. Scott.*

Captain James T. Kirk

*A coin is spinning. The coin is in a strong box sitting in the mud at the bottom of the deepest ocean trench. Don't ask what has set the coin spinning or what is keeping it spinning. This isn't a well-thought-out story! The point is that there is an identical spinning coin in an identical box sitting on a cold moon in a distant galaxy on the far side of the Universe.*

*The first coin comes down heads. Instantaneously, without the merest split-second of delay, its cousin 10 billion light-years from Earth comes down tails.*

The coin on Earth could equally well have come down tails and its distant cousin heads. This is not important. The significant thing is that the coin on the far side of the Universe *knows* instantly the state of its distant terrestrial cousin—and does the opposite.

But how can it possibly know? The cosmic speed limit in our Universe is the speed of light.[1] Since the coins are separated by 10 billion light-years, information about the state of one coin must take

---

[1]See Chapter 7, "The Death of Space and Time."

a minimum of 10 billion years to reach the other. Yet they know about each other in a split second.

This kind of "spooky action at a distance" turns out to be one of the most remarkable features of the microscopic world. It so upset Einstein that he declared that quantum theory must be wrong. In fact, Einstein was wrong.

In the past 20 years, physicists have observed the behaviour of coins that are separated by large distances. The coins are quantum coins, and the distances are not of course as large as the width of the Universe.[2] Nevertheless, the experiments have successfully demonstrated that atoms and their kin can indeed communicate instantaneously, in total violation of the speed-of-light barrier.

Physicists have christened this weird kind of quantum telepathy nonlocality. The best way to understand it is by considering a peculiar particle property called spin.

## SPOOKY ACTION AT A DISTANCE

Spin is unique to the microscopic world. Particles that possess spin behave as if they are rotating like tiny spinning tops. Only they aren't actually spinning! Once again, we come up against the fundamental ungraspability of the microscopic world. The spin of particles, like their inherent unpredictability, is something with no direct analogue in the everyday world. Microscopic particles can have different amounts of spin. The electron happens to carry the minimum quan-

---

[2]In fact, the quantum coins have to be created together, then separated, to show spooky action at a distance, which is another reason the tale of coins on different sides of the Universe shouldn't be taken too seriously. As pointed out, it isn't a well-thought-out story. It exists merely to convey one amazing truth and one amazing truth only—that quantum theory permits objects to influence each other instantaneously, even when on opposite sides of the Universe.

tity. This permits it to spin in two possible ways. Think of it as spinning either clockwise or anticlockwise (although of course it isn't
actually spinning at all!).

If two electrons are created together—the first with clockwise
spin, the second with anticlockwise spin—their spins cancel. Physicists say their total spin is zero. Of course, the pair of electrons can
also have a total spin of zero if the first electron has an anticlockwise
spin and the second a clockwise spin.

Now, there is a law of nature that says the total spin of such a
system can never change. (It's actually called the law of conservation
of angular momentum.) So once the pair of electrons has been created with a total spin of zero, the pair's spin must remain zero as long
as the pair remains in existence.

Nothing out of the ordinary here. However, there is another way
to create two electrons with a total spin of zero. Recall that, if two
states of a microscopic system are possible, then a superposition of
the two is also possible. This means it is possible to create a pair of
electrons that are simultaneously clockwise-anticlockwise and
anticlockwise-clockwise.

So what? Well, remember that such a superposition can exist only
as long as the pair of electrons is isolated from its environment. The
moment the outside world interacts with it—and that interaction
could be someone checking to see what the electrons are doing—the
superposition undergoes decoherence and is destroyed. Unable to exist any longer in their schizophrenic state, the electrons plump for
being either clockwise-anticlockwise or anticlockwise-clockwise.

Still nothing out of the ordinary (at least for the microscopic
world!). However, imagine that, after the electrons are created in their
schizophrenic state, they remain isolated and nobody looks at them.
Instead, one electron is taken away in a box to a faraway place. Only
then does someone finally open the box and observe the spin of the
electron.

If the electron at the faraway place turns out to have a clockwise
spin, then instantaneously the other electron must stop being in its

schizophrenic state and assume an anticlockwise spin. The total spin, after all, must always remain zero. If, on the other hand, the electron turns out to be spinning anticlockwise, its cousin must instantaneously assume a clockwise spin.

It does not matter if one electron is in a steel box half-buried on the seafloor and the other is in a box on the far side of the Universe. One electron will respond instantaneously to the other's state. This is not merely some esoteric theory. Instantaneous influence has actually been observed in the laboratory.

In 1982, Alain Aspect and his colleagues at the University of Paris South created pairs of photons and sent members of each pair to detectors separated by a distance of 13 metres. The detectors measured the polarisation of the photons, a property related to their spin. Aspect's team showed that measuring the polarisation of photons at one detector affected the polarisation measured at the other detector. The influence that travelled between the detectors did so in less than 10 nanoseconds. Crucially, this was a quarter of the time a light beam would have taken to bridge the 13-metre gap.

At the bare minimum, some kind of influence travelled between the detectors at four times the speed of light. If the technology had made it possible to measure an even smaller time interval, Aspect could have shown the ghostly influence to be even faster. Quantum theory was right. And Einstein—bless him—was wrong.

Nonlocality could never happen in the ordinary, nonquantum world. An air mass might split into two tornadoes, one spinning clockwise and the other anticlockwise. But that's the way they would stay—spinning in opposite directions—until finally they both ran out of steam. The crucial difference in the microscopic, quantum world is that the spins of particles are undetermined until the instant they are observed. And, before the spin of one electron in the pair is observed, it is totally unpredictable. It has a 50 per cent chance of being clockwise and a 50 per cent chance of being anticlockwise (once again we come up against the naked randomness of the microworld). But even though there is no way of knowing the spin of one electron until it is observed, the spin of the other electron must settle down to being opposite instantaneously—no matter how far away the other particle happens to be.

## ENTANGLEMENT

At the heart of nonlocality is the tendency of particles that interact with each other to become entwined, or "entangled", so that the properties of one are forever dependent on the properties of the other. In the case of the pair of electrons, it is their spins that become dependent on each other. In a very real sense, entangled particles cease to have a separate existence. Like a much-in-love couple, they become a weird joined-at-the-hip entity. No matter how far apart they are pulled, they remain forever connected.

The weirdest manifestation of entanglement is, without doubt, nonlocality. In fact, it would seem that if we could harness it we could create an instantaneous communications system. With it we could phone the other side of the world with no time delay. In fact, we could phone the other side of the Universe with no time delay! No longer would we need to be inconvenienced by the pesky speed-of-light barrier.

Frustratingly, however, nonlocality cannot be harnessed to create an instantaneous communications system. Attempts to use the spin of particles to send a message across large distances might use one direction of spin to code for a "0" and the other for a "1." However, to know that you were sending a "0" or a "1," you would have to check the spin of the particle. But checking kills the superposition, which is essential for the instantaneous effect. If you sent a message without first looking, you could be only 50 per cent sure of sending a "1," a level of uncertainty that effectively scrambles any meaningful message.

So although instantaneous influence is a fundamental feature of our Universe, it turns out that nature does exactly what is required to make it unusable for sending real information. This is how it permits

the speed-of-light barrier to be broken without actually being broken. What nature gives with one hand it cruelly takes away with the
other.

## TELEPORTATION

Arguably, the sexiest potential use of entanglement involves taking an
object and sending a complete description of the object to a faraway
place so that a suitably clever machine at'the other end can construct
a perfect copy. This is of course the recipe for the *Star Trek* transporter, which routinely "beamed" crew members back and forth between planet and ship.

The technology to reconstruct a solid object merely from the information describing it is of course way beyond our current techno-
logical capabilities. But, actually, the idea of creating a perfect copy of
an object at a remote location founders on something much more
basic than this. According to the Heisenberg uncertainty principle, it
is impossible to perfectly describe an object—the positions of all its
atoms, the electrons in each of those atoms, and so on. Without such
knowledge, however, how can an exact copy ever be assembled?

Entanglement, remarkably, offers a way out. The reason is that
entangled particles behave like a single indivisible entity. At some
level, they *know* each other's deepest secrets.

Say we have a particle, P, and we want to make a perfect copy, P\*.
It stands to reason that in order to do this it is necessary to know P's
properties. However, according to the Heisenberg uncertainty principle, if we measure one particular property of P precisely—say its
location—we inevitably lose all knowledge of some other property—
in this case, its velocity. Nevertheless, this crippling limitation can be
circumvented by an ingenious use of entanglement.

Take another particle, A, which is similar to both P and P\*. The
important thing is that A and P\* are an entangled pair. Now, entangle
A with P and make a measurement of the pair together. This will tell
us about some property of P. According to the Heisenberg uncer-

tainty principle, however, the measurement will inevitably involve us losing knowledge of some other property of P.

But all is not lost. Because P* was entangled with A, it retains knowledge about A. And because A was entangled with P, it retains knowledge about P. This means that P*, though it has never been in touch with P, nevertheless knows its secrets. Furthermore, when the measurement was made on A and P together and information about some property of P seemed to be lost, instantaneously it became available to A's partner, P*. This is the miracle of entanglement.

Since we already know about the other properties of P, obtained from A, we now have all we need to make sure P* has *exactly* the attributes of P.[3] Thus we have exploited entanglement to circumvent the restrictions of the Heisenberg uncertainty principle.

The amazing thing is that, although we have exploited entanglement to make a particle P* with the exact properties of P, at no time did we ever possess any information about the missing property of P! It was transmitted out of our sight through the ghostly connections of entanglement.[4]

Calling this scheme teleportation is a bit of a cheeky exaggeration since it solves only one of the many problems in making a *Star Trek* transporter. The researchers of course knew this. But they also knew a thing or two about how to grab newspaper headlines!

As it happens, the Achilles' heel of the *Star Trek* transporter turns out to be neither pinning down the position, and so on, of every atom

---

[3]The information on the original particle, P, must be transmitted by ordinary means—that is, slower than the speed of light, the cosmos's speed limit. So even if P and P* are far apart, the creation of P*—the perfect copy of P—is not instantaneous, despite the fact that communication between the entangled particles, A and P, is instantaneous.

[4]It is worth emphasising that, even with entanglement, the most you can ever do is make a copy of an object at the expense of destroying the original. Making a copy and keeping the original is impossible.

in a person's body nor assembling a copy of the person from that information. It's actually *transmitting* the sheer volume of information needed to describe a person across space. Billions of times more information is needed than for the reconstruction of a two-dimensional TV image. The obvious way to send the information is as a series of binary "bits"—dots and dashes. If the information is to be sent in a reasonable time, the pulses must obviously be short. But ultrashort pulses are possible only with ultra-high-energy light. As science fiction writer Arthur C. Clarke has pointed out, beaming up Captain Kirk could easily take more energy than there is in a small galaxy of stars!

Teleportation and nonlocality aside, the most mind-blowing consequence of entanglement is what it means for the Universe as a whole. At one time, all particles in the Universe were in the same state because all particles were together in the Big Bang. Consequently, all particles in the Universe are to some extent entangled with each other.

There is a ghostly web of quantum connections crisscrossing the Universe and coupling you and me to every last bit of matter in the most distant galaxy. We live in a telepathic universe. What this actually means physicists have not yet figured out.

Entanglement may also help explain the outstanding question posed by quantum theory: Where does the everyday world come from?

## WHERE DOES THE EVERYDAY WORLD COME FROM?

According to quantum theory, weird superpositions of states are not only possible but guaranteed. An atom can be in many places at once or do many things at once. It is the interference between these possibilities that leads directly to many of the bizarre phenomena observed in the microscopic world. But why is it that, when large numbers of atoms club together to form everyday objects, those objects never display quantum behaviour? For instance, trees never behave as if they are in two places at once and no animal behaves as if it is a combination of a frog and a giraffe.

The first attempt to explain the conundrum was made in Copenhagen in the 1920s by quantum pioneer Niels Bohr. The Copenhagen Interpretation, in effect, divides the Universe into two domains, ruled by different laws. On the one hand, there is the domain of the very small, which is ruled by quantum theory, and on the other there is the domain of the very big, ruled by normal, or classical, laws. According to the Copenhagen Interpretation, it is when a quantum object like an atom interacts with a classical object that it is forced to stop being in a schizophrenic superposition and start behaving sensibly. The classical object could be a detecting device or even a human being.

But what exactly does a classical object do to stop a quantum object from being quantum? Even more importantly, what constitutes a classical object? After all, an eye is just a big collection of atoms, which individually obey quantum theory. This turns out to be the Achilles' heel of the Copenhagen Interpretation and the reason it has always appeared to many to be a very unsatisfactory explanation of where the everyday world comes from.

The Copenhagen Interpretation divides the universe, arbitrarily, into two domains, only one of which is governed by quantum theory. This in itself is very defeatist. After all, if quantum theory is a fundamental description of reality, surely it should apply everywhere—to the atomic world and the everyday world. The idea that it is a universal theory is, in a nutshell, what physicists believe today.

It turns out we never observe a quantum system directly. We only observe its effect on its environment. This may be a measuring device or a human eye or, in general, the universe. For instance, the light from an object impinges on the retina of the eye and makes an impression there. What the observer *knows* is inseparable from what the observer *is*. Now, if quantum theory applies everywhere, we have a quantum object observing, or recording, another quantum object. The central question can therefore be restated: Why do weird schizophrenic states fail to impress themselves on, or *entangle* themselves with, the environment, whereas everyday one-place-at-one-time states do? An example may help.

If a high-speed subatomic particle flies through the air, it knocks electrons from any atoms it passes close to. Imagine it was possible to see a 10-centimetre-long portion of its track. And, say in that 10 centimetres the particle has a 50 per cent chance of interacting with one electron, kicking it out of its parent atom.

The particle, therefore, either knocks out an electron or doesn't knock out an electron. But because the event of knocking out an electron is a quantum event, there is another possibility—the superposition of the two events. The particle both knocks out an electron and doesn't knock out an electron! The question is: Why, when this event entangles itself with the environment, does it not leave an indelible impression? As luck would have it, it is possible to actually *see* an electron ejection event with an ingenious device known as a cloud chamber.

Clouds form in the air when a drop in temperature causes water droplets to condense out of water vapour. But this process happens rapidly only if there are things like dust particles in the air that act as "seeds" around which water droplets can grow. Now the seed—and this is the key to the cloud chamber's operation—need not be as big as a dust grain. In fact, it need be only a single atom that has lost an electron—an ion.

A cloud chamber is a box filled with water vapour with a window in its side to look through. Crucially, the water vapour is ultrapure, so there are no seeds about which the vapour can condense. The vapour is held in a state in which it is absolutely desperate to form droplets, but it is frustrated because there are no seeds. Enter a high-speed subatomic particle. Where it knocks an electron out of an atom, a water droplet will instantly grow around the ion. The droplet is small but big enough to see through the window of the cloud chamber if properly illuminated.

So what would you see if you looked through the window? The

answer is of course just one of the possibilities—either a single water droplet or no water droplet. You would never see a superposition of both—a ghostly droplet, hovering half in existence and half out of existence. The question is, what happens in the cloud chamber to prevent it from recording this superposition?

Take the event in which a water droplet forms. It was triggered by a single ionised atom. The same atom exists in the event in which no droplet formed. It just does not get ionised, so no water droplet forms around it. Say, this atom is painted red in both cases to make it stand out (forget the fact that you can't paint an atom!).

Now, in the event a droplet forms, zoom in on an atom near the red atom. Water is denser than water vapour; the atoms are closer together. Consequently, the atom in question will be closer to the red atom than it is in the event in which no water droplet forms. For this reason, the probability wave representing the atom in the first event only partially overlaps with the probability wave of the same atom in the second event. Say, for example, that their waves only half overlap.

Now take a second atom in the first event. It too will be closer in the first case than in the second. Once again, their probability waves will only half overlap. If we now consider the probability wave representing the two atoms together, it will overlap only one-quarter with the second case, since $^1/_2 \times {}^1/_2 = {}^1/_4$.

See where this is going? Say the water droplet contains a million atoms, which actually corresponds to a very small droplet. How much will the probability wave representing a million atoms in the first event overlap with the probability wave representing a million atoms in the second event? The answer is $^1/_2 \times {}^1/_2 \times {}^1/_2 \times \ldots$ a million times. This is an extraordinarily small number. There will therefore be essentially zero overlap.

But if two waves don't overlap at all, how can they interfere? The answer is, of course, they cannot. Interference, however, is at the root of all quantum phenomena. If interference between the two events is impossible, we see either one event or the other but never the effect of one event mingling with the other, the essence of quantumness.

Probability waves that do not overlap and so cannot interfere are said to have lost coherence, or to have *decohered*. Decoherence is the ultimate reason why the record of a quantum event in the environment, which always consists of a lot of atoms, is never quantum. In the case of the cloud chamber, the "environment" is the million atoms around the ionised/nonionised atom. In general, however, the environment consists of the countless quadrillions of atoms in the Universe. Decoherence is therefore hugely effective at destroying any overlap between the probability waves of events entangled with the environment. And since that's the only way we can experience them— what the observer *knows* is inseparable from what the observer *is*— we never directly see quantum behaviour.

# 6

# IDENTICALNESS AND THE
# ROOTS OF DIVERSITY

HOW THE BEWILDERING VARIETY OF THE EVERYDAY WORLD STEMS FROM THE
FACT THAT YOU CANNOT TATTOO AN ELECTRON

*I woke up one morning and all of my stuff had been stolen, and re-
placed by exact duplicates.*

Steven Wright

*They came from far and wide to see it—the river that ran uphill. It
flowed past the fishing port, climbed through the close-packed houses,
before meandering up the sheep-strewn hillside to the craggy summit
overlooking the town. Startled seagulls bobbed on it. Excited children
ran beside it. And at picnic tables outside pubs all along the river's lower
reaches, daytrippers sat transfixed by this wonder of nature as beer crept
steadily up the sides of their beer glasses and quietly emptied itself onto
the ground.*

Surely, there is no liquid that can defy gravity like this and run uphill?
Remarkably, there is. It's yet another consequence of quantum theory.

Atoms and their kin can do many impossible things before break-
fast. For instance, they can be in two or more places at once, penetrate
impenetrable barriers, and *know* about each other instantly even
when on different sides of the Universe. They are also totally unpre-

dictable, doing things for no reason at all—perhaps the most shocking and unsettling of all their characteristics.

All of these phenomena ultimately come down to the wave-particle character of electrons, photons, and their like. But the peculiar dual nature of microscopic objects is not the only thing that makes them profoundly different from everyday objects. There is something else: their *indistinguishability*. Every electron is identical to every other electron, every photon is identical to every other photon, and so on.[1]

At first sight this may not seem a very remarkable property. But think of objects in the everyday world. Although two cars of the same model and colour appear the same, in reality they are not. A careful inspection would reveal that they differ slightly in the evenness of their paint, in the air pressure in their tires, and in a thousand other minor ways.

Contrast this with the world of the very small. Microscopic particles cannot be scratched or marked in any way. You cannot tattoo an electron! They are utterly indistinguishable.[2] The same is true of photons and all other denizens of the microscopic world. This indistinguishability is truly something new under the Sun. And it has remarkable consequences for both the microscopic world and the everyday world. In fact, it is fair to say that it is the reason the world we live in is possible.

---

[1] Since photons come with different *wavelengths,* we are of course talking here about photons *with the same wavelength* being identical to each other.

[2] John Wheeler and Richard Feynman once came up with an interesting suggestion for why electrons are utterly indistinguishable—because there is only one electron in the Universe! It weaves backwards and forwards in time like a thread going back and forth through a tapestry. We see the multitude of places where the thread goes through the fabric of the tapestry and mistakenly attribute each to a separate electron.

### THINGS YOU CAN'T TELL APART INTERFERE

Recall that all the bizarre behaviour in the microscopic world, such as an atom's ability to be in many places at once, comes down to interference. In the double slit experiment, for example, it is the interference between the wave corresponding to a particle going through the left-hand slit and the wave corresponding to the particle going through the right-hand slit that produces the characteristic pattern of alternating dark and light stripes on the second screen.

Recall also that if you were to set up some means of determining which slit each particle goes through—enabling you to distinguish between the two alternative events—the interference stripes disappear because of decoherence. Interference, it turns out, happens only if the alternative events are *indistinguishable*—in this case, the particle going through one slit and the particle going through the other slit.

In the case of the double slit experiment, the alternative events are indistinguishable just as long as nobody looks. But identical particles, such as electrons, raise the possibility of entirely new kinds of indistinguishable events.

Think of a teenage boy who plans to go out clubbing with his girlfriend, who happens to have an identical twin sister. Unbeknown to him, his girlfriend decides to stay in and watch TV and sends her twin in her place. Because the two girls appear identical to the boy (although they are not of course identical at the microscopic level), the events of going clubbing with his girlfriend and going clubbing with his girlfriend's sister are indistinguishable.

Events such as this one, which are indistinguishable simply because they involve apparently indistinguishable things, have no serious consequences in the wider world (apart from allowing identical twin girls to run rings around their boyfriends). However, in the microscopic world, they have truly profound consequences. Why? Because events that are indistinguishable—for any reason whatsoever—are able to interfere with each other.

## THE COLLISION OF IDENTICAL THINGS

Take two atomic nuclei that collide. Any such collision—and this particular point will have to be taken on trust—can be looked at from a point of view in which the nuclei fly in from opposite directions, hit, then fly back out in opposite directions. In general, the in and out directions are not the same. Think of a clock face. If the nuclei fly into the collision point from, say, 9:00 and 3:00, they might fly out toward 4:00 and 10:00. Or 1:00 and 7:00. Or any other pair of directions, as long as the directions are opposite each other.

An experimenter could tell which direction the two nuclei ricochet by placing detectors at opposite sides of the imaginary clock face and then moving them around the rim together. Say the detectors are placed at 4:00 and 10:00. In this case, there are two possible ways the nuclei can get to the detectors. They could strike each other with a glancing blow so that the one coming from 9:00 hits the detector at 4:00 and the one coming from 3:00 hits the one at 10:00. Or they could hit head on, so that the one coming from 9:00 bounces back almost the way it came and hits the detector at 10:00 and the one coming from 3:00 bounces back almost the way it came and hits the detector at 4:00.

The directions of 4:00 and 10:00 are in no way special. Wherever the two detectors are positioned, there will be two alternative ways the nuclei can get to them. Call them events A and B.

What happens if the two nuclei are different? Say the one that flies in from 9:00 is a nucleus of carbon and the one that flies in from 3:00 is a nucleus of helium. Well, in this case, it is always possible to distinguish between events A and B. After all, if a carbon nucleus is picked up by the detector at 10:00, it is obvious that event A occurred; if it is picked up by the detector at 3:00, it must have been event B instead.

What happens, however, if the two nuclei are the same? Say each is a nucleus of helium? Well, in this case, it is impossible to distinguish between events A and B. A helium nucleus that is picked up by the detector in the direction of 10:00 could have got there by either route, and the same is true for a helium nucleus picked up in the direction of 4:00. Events A and B are now indistinguishable. And if two events in the microscopic world are indistinguishable, the waves associated with them interfere.

In the collision of two nuclei, interference makes a huge difference. For instance, it is possible that the two waves associated with the two indistinguishable collision events destructively interfere, or cancel each other out, in the direction of 10:00 and 4:00. If so the detectors will pick up no nuclei at all, no matter how many times the experiment is repeated. It is also possible that the two waves constructively interfere, or reinforce each other, in the direction of 10:00 and 4:00. In this case, the detectors will pick up an unusually large number of nuclei.

In general, because of interference, there will be certain outward directions in which the waves corresponding to events A and B cancel each other and certain outward directions in which they reinforce each other. So if the experiment is repeated thousands of times and the ricocheting nuclei are picked up by detectors all around the rim of the imaginary clock face, the detectors will see a tremendous variation in the number of nuclei arriving. Some detectors will pick up many nuclei, while others will pick up none at all.

But this is dramatically different from the case when the nuclei are different. Then there is no interference, and the detectors will pick up nuclei ricocheting in all directions. There will be no places around the clock face where nuclei are not seen.

This striking difference between the outcomes of the experiment when the nuclei are the same and when they are different is not because of the difference in masses of the nuclei of carbon and helium, although this has a small effect. It is truly down to whether collision events A and B are distinguishable or not.

If this kind of thing happened in the real world, think what it would mean. A red bowling ball and a blue bowling ball that are repeatedly collided together would ricochet in all possible directions.

But everything would be changed merely by painting the red ball blue so the two balls were indistinguishable. Suddenly, there would be directions in which the balls ricocheted far more often than when they were different colours and directions in which they never, ever ricocheted.

This fact, that events involving identical particles in the microscopic world can interfere with each other, may seem little more than a quantum quirk. But it isn't. It is the reason why there are 92 different kinds of naturally occurring atoms rather than just 1. In short, it is responsible for the variety of the world we live in. Understanding why, however, requires appreciating one more subtlety of the process in which identical particles collide.

## TWO TRIBES OF PARTICLES

Recall the case in which the nuclei are different—a carbon nucleus and a helium nucleus—and consider again the two possible collision events. In one, the nuclei strike each other with a glancing blow, and in the other they hit head on and bounce back almost the way they came. What this means is that, for the nucleus that comes in at 9:00, there is a wave corresponding to it going out at 4:00 and a wave corresponding to it going out at 10:00.

The key thing to understand here is that the probability of an event is not related to the height of the wave associated with that event but to the square of the height of the wave. The probability of the 4:00 event is therefore the square of the wave height in the direction of 4:00 and the probability of the 10:00 event is the square of the wave height in the direction of 10:00. It is here that the crucial subtlety comes in.

Say the wave corresponding to the nucleus that flies out at 10:00 is flipped by the collision, so that its troughs become its peaks and its peaks become its troughs. Would it make any difference to the probability of the event? To answer this, consider a water wave—a series of alternating peaks and troughs. Think of the average level of the water as corresponding to a height equal to zero so that the height of the peaks is a positive number—say plus 1 metre—and the height of the troughs is a negative number—minus 1 metre. Now it makes no difference whether you square the height of a peak or the height of a trough since $1 \times 1 = 1$ and $-1 \times -1$ also equals 1. Consequently, flipping the probability wave associated with a ricocheting nucleus makes no difference to the event's probability.

But is there any reason to believe that one wave might get flipped? Well, the 10:00 collision and the 4:00 collision are very different events. In one, the trajectory of the nucleus hardly changes whereas in the other it is turned violently back on itself. It is at least plausible that the 10:00 wave might get flipped.

Just because something is plausible does not mean it actually happens. True. In this case, however, it does! Nature has two possibilities available to it: It can flip the wave of one collision event or it can leave it alone. It turns out that it avails itself of both.

But how would we ever know about a probability wave getting flipped? After all, the only thing an experimenter can measure is the number of nuclei picked up by a detector which depends on the probability of a particular collision event. But this is determined by the square of the wave height, which is the same whether the wave is flipped or not. It would seem that what actually happens to the probability wave in the collision is hidden from view.

If the colliding particles are different, this is certainly true. But, crucially, it is not if they are identical. The reason is that the waves corresponding to indistinguishable events interfere with each other. And in interference it matters tremendously whether or not a wave is flipped before it combines with another wave. It could mean the difference between peaks and troughs coinciding or not, between the waves cancelling or boosting each other.

What happens then in the collision of identical particles? Well, this is the peculiar thing. For some particles—for instance, photons—everything is the same as it is for identical helium nuclei. The waves that correspond to the two alternative collision events interfere with each other normally. However, for other particles—for instance, electrons—things are radically different. The waves corresponding to the two alternative collision events interfere, but only after one has been flipped.

Nature's basic building blocks turn out to be divided into two tribes. On the one hand, there are particles whose waves interfere with each other in the normal way. These are known as bosons. They include photons and gravitons, the hypothetical carriers of the gravitational force. And, on the other hand, there are particles whose waves interfere with one wave flipped. These are known as fermions. They include electrons, neutrinos, and muons.

Whether particles are fermions or bosons—that is, whether or not they indulge in waveflipping—turns out to depend on their spin. Recall that particles with more spin than others behave as if they are spinning faster about their axis (although in the bizarre quantum world particles that possess spin are not actually spinning!). Well, it turns out that there is a basic indivisible chunk of spin, just like there is a basic indivisible chunk of everything in the microscopic world. For historic reasons, this "quantum" of spin is $^1/_2$ unit (don't worry what the units are). Bosons have integer spin—0 units, 1 unit, 2 units, and so on—and fermions have "half-integer" spin—$^1/_2$ unit, $^3/_2$ units, $^5/_2$ units, and so on.

Why do particles with half-integer spin indulge in waveflipping, whereas particles with integer spin do not? This, of course, is a very good question. But it brings us to the end of what can easily be conveyed without opaque mathematics. Richard Feynman at least came clean about this: "This seems to be one of the few places in physics where there is a rule which can be stated very simply but for which no one has found an easy explanation. It probably means that we do not have a complete understanding of the fundamental principles involved."

Feynman, who worked on the atomic bomb and won the 1965 Nobel Prize for Physics, was arguably the greatest physicist of the postwar era. If you find the ideas of quantum theory a little difficult, you are therefore in very good company. It is fair to say that, 80-odd years after the birth of quantum theory, physicists are still waiting for the fog to lift so that they can clearly see what it is trying to tell us about fundamental reality. As Feynman himself said: "I think I can safely say that nobody understands quantum mechanics."

Brushing the spin mystery under the carpet, we come finally to the point of all this—the implication of waveflipping for fermions such as electrons.

Instead of two helium nuclei, think of two electrons, each of which collides with another particle. After the collision, they ricochet in almost the same direction. Call the electrons A and B and call the directions 1 and 2 (even though they are almost the same direction). Exactly as in the case of two identical nuclei, there are two indistinguishable possibilities. Electron A could ricochet in direction 1 and electron B in direction 2, or electron A could ricochet in direction 2 and electron B in direction 1.

Since electrons are fermions, the wave corresponding to one possibility will be flipped before it interferes with the wave corresponding to the other possibility. Crucially, however, the waves for the two possibilities are identical, or pretty identical. After all, we are talking about two identical particles doing almost identical things. But if you add two identical waves—one of which has been flipped—the peaks of one will exactly match the troughs of the other. They will completely cancel each other out. In other words, the probability of two electrons ricocheting in exactly the same direction is zero. It is completely impossible!

This result is actually far more general than it appears. It turns out that two electrons are not only forbidden from ricocheting in the same direction, they are forbidden from doing the same thing, period. This prohibition, known as the Pauli exclusion principle, after Austrian physicist Wolfgang Pauli, turns out to be the ultimate reason for the existence of white dwarfs. While it is certainly true that an electron cannot be confined in too small a volume of space, this still does not explain why all the electrons in a white dwarf do not simply crowd together in exactly the same small volume. The Pauli exclusion principle provides the answer. Two electrons cannot be in the same quantum state. Electrons are hugely antisocial and avoid each other like the plague.

Think of it this way. Because of the Heisenberg uncertainty principle, there is a minimum-sized "box" in which an electron can be squeezed by the gravity of a white dwarf. However, because of the Pauli exclusion principle, each electron demands a box to itself. These two effects, working in concert, give an apparently flimsy gas of electrons the necessary "stiffness" to resist being squeezed by a white dwarf's immense gravity.

Actually, there is yet another subtlety here. The Pauli exclusion principle prevents two fermions from doing the same thing if they are identical. But electrons have a way of being different from each other because of their spin. One can behave as if it is spinning clockwise and one as if it is spinning anticlockwise.[3] Because of this property of electrons, *two* electrons are permitted to occupy the same volume of space. They may be unsociable, but they are not complete loners! White dwarfs are hardly everyday objects. However, the Pauli exclusion principle has much more mundane implications. In particular, it explains why there are so many different types of atoms and why the world around us is the complex and interesting place it is.

## WHY ATOMS AREN'T ALL THE SAME

Recall that, just as sound waves confined in an organ pipe can vibrate in only restricted ways, so too can the waves associated with an electron confined in an atom. Each distinct vibration corresponds to a possible orbit for an electron at a particular distance from the central nucleus and with a particular energy. (Actually, of course, the orbit is

---

[3]Physicists call two alternatives spin "up" and spin "down." But that is just a technicality.

merely the most probable place to find an electron since there is no such thing as a 100 per cent certain path for an electron or any other microscopic particle.)

Physicists and chemists number the orbits. The innermost orbit, also known as the ground state, is numbered 1, and orbits successively more distant from the nucleus are numbered 2, 3, 4, and so on. The existence of these quantum numbers, as they are called, emphasises yet again how everything in the microscopic world—even the orbits of electrons—comes in discrete steps with no possibility of intermediate values.

Whenever an electron "jumps" from one orbit to another orbit closer to the nucleus, the atom loses energy, which is given out in the form of a photon of light. The energy of the photon is exactly equal to the difference in energy of two orbits. The opposite process involves an atom absorbing a photon with an energy equal to the difference in energy of two orbits. In this case, an electron jumps from one orbit to another orbit farther from the nucleus.

This picture of the "emission" and "absorption" of light explains why photons of only special energies—corresponding to special frequencies—are spat out and swallowed by each kind of atom. The special energies are simply the energy differences between the electron orbits. It is because there is a limited number of permitted orbits that there is a restricted number of orbital "transitions."

But things are not quite this simple. The electron waves that are permitted to vibrate inside an atom can be quite complex three-dimensional things. They may correspond to an electron that is not only most likely to be found at a certain distance from the nucleus but more likely to be found in some directions rather than others. For instance, an electron wave might be bigger over the north and south poles of the atom than in other directions. An electron in such an orbit would most likely be found over the north and south poles.

Describing a direction in three-dimensional space requires two numbers. Think of a terrestrial globe where a latitude and longitude are required. Similarly, in addition to the numbers specifying its distance from the nucleus, an electron wave whose height changes with direction requires two more quantum numbers to describe it. This makes a total of three. In recognition of the fact that electron orbits are totally unlike more familiar orbits—for instance, the orbits of planets around the Sun—they are given a special name: orbitals.

The precise shape of electron orbitals turns out to be crucially important in determining how different atoms stick together to make molecules such as water and carbon dioxide. Here, the key electrons are the outermost ones. For instance, an outer electron from one atom might be shared with another atom, creating a chemical bond. Where exactly the outermost electron is clearly plays an important role. If, for example, it has its highest probability of being found above the atom's north and south poles, the atom will most easily bond with atoms to its north or south.

The science that concerns itself with all the myriad ways in which atoms can join together is chemistry. Atoms are the ultimate Lego bricks. By combining them in different ways, it is possible to make a rose or a gold bar or a human being. But exactly how the Lego bricks combine to create the bewildering variety of objects we see in the world around us is determined by quantum theory.

Of course, an obvious requirement for the existence of a large number of combinations of Lego bricks is that there be more than one kind of brick. Nature in fact uses 92 Lego bricks. They range from hydrogen, the lightest naturally occurring atom, to uranium, the heaviest. But why are there so many different atoms? Why are they not all the same? Once again, it all comes down to quantum theory.

## WHY ATOMS ARE NOT ALL THE SAME

Electrons trapped in the electric force field of a nucleus are like foot-balls trapped in a steep valley. By rights they should run rapidly down-hill to the lowest possible place—the innermost orbital. But if this was what the electrons in atoms really did, all atoms would be roughly the same size. More seriously, since the outermost electrons determine how an atom bonds, all atoms would bond in exactly the same way. Nature would have only one kind of Lego brick to play with and the world would be a very dull place indeed.

What rescues the world from being a dull place is the Pauli exclusion principle. If electrons were bosons, it is certainly true that an atom's electrons would all pile on top of each other in the innermost orbital. But electrons are not bosons. They are fermions. And fermions abhor being crowded together.

This is how it works. Different kinds of atoms have different numbers of electrons (always of course balanced by an equal number of protons in their nuclei). For instance, the lightest atom, hydrogen, has one electron and the heaviest naturally occurring atom, uranium, has 92. In this discussion the nucleus is not important. Focus instead on the electrons. Imagine starting with a hydrogen atom and then adding electrons, one at a time.

The first available orbit is the innermost one, nearest the nucleus. As electrons are added, they will first go into this orbit. When it is full and can take no more electrons, they will pile into the next available orbit, farther away from the nucleus. Once that orbit is full, they will fill the next most distant one. And so on.

All the orbitals at a particular distance from the nucleus—that is, with different directional quantum numbers—are said to make up a shell. The maximum number of electrons that can occupy the innermost shell turns out to be two—one electron with clockwise spin and one with anticlockwise spin. A hydrogen atom has one electron in this shell and an atom of helium, the next biggest atom, has two.

The next biggest atom is lithium. It has three electrons. Since there is no more room in the innermost shell, the third electron starts a new shell farther out from the nucleus. The capacity of this shell is eight. For atoms with more than 10 electrons, even this shell is all used up, and another begins to fill up yet farther from the nucleus.

The Pauli exclusion principle, by forbidding more than two electrons from being in the same orbital—that is, from having the same quantum numbers—is the reason that atoms are different from each other. It is also responsible for the rigidity of matter. "It is the fact that electrons cannot get on top of each other that makes tables and everything else solid," said Richard Feynman.

Since the manner in which an atom behaves—its very identity—depends on its outer electrons, atoms with similar numbers of electrons in their outermost shells tend to behave in a similar way. Lithium, with three electrons, has one electron in its outer shell. So too does sodium, with 11 electrons. Lithium and sodium therefore bond with similar kinds of atoms and have similar properties.

So much for fermions, which are subject to the Pauli exclusion principle. What about bosons? Well, since such particles are not governed by the exclusion principle, they are positively gregarious. And this gregariousness leads to a host of remarkable phenomena, from lasers to electrical currents that flow forever to liquids that flow uphill.

## WHY BOSONS LIKE TO BE TOGETHER WITH THEIR MATES

Say two boson particles fly into a small region of space. One hits an obstruction in its path and ricochets; the other hits a second obstruction and ricochets. It doesn't matter what the obstructing bodies are; they may be nuclei or anything else. The important thing here is the direction in which they ricochet, which is the same for both.

Call the particles A and B, and call the directions they ricochet in 1 and 2 (even if they are almost the same direction!). There are two possibilities. One is that particle A ends up in direction 1 and particle B ends up in direction 2. The other is that A ends up in direction 2 and B in direction 1. Because A and B are schizophrenic denizens of the microscopic world, there is a wave corresponding to A going in direction 1 and to B in direction 2. And there is also a wave corresponding to A going in direction 2 and to B in direction 1.

If the two bosons are different particles there can be no interference between them. So the probability that a detector picks up the two ricocheting particles is simply the square of the height of the first wave plus the square of the height of the second wave, since the probability of anything happening in the microscopic world is always the square of the height of the wave associated with it. Well, it turns out—and this will have to be taken on trust—that the two probabilities are roughly the same. So the overall probability simply is twice the probability of each event happening individually.

Say the waves have a height of 1 for both processes. This would mean that if they were squared and added to get the probability for both processes, it would be $(1 \times 1) + (1 \times 1) = 2$. Now a probability of 1 corresponds to 100 per cent, so a probability of 2 is clearly ridiculous! But bear with this. It is still possible to make a comparison of probabilities, which is where all this is leading.

Now, say the two bosons are identical particles. In this case, the two possibilities—A going in direction 1 and B in direction 2, and A going in direction 2 and B in direction 1—are indistinguishable. And because they are indistinguishable, the waves associated with them can interfere with each other. Their combined height is $1 + 1$. The probability for both processes is therefore $(1 + 1) \times (1 + 1) = 4$.

This is twice as big as when the bosons were not identical. In other words, if two bosons are identical, they are twice as likely to ricochet in the same direction as if they were different. Or to put it another way, a boson is twice as likely to ricochet in a particular direction if another boson ricochets in that direction too.

The more bosons there are the more significant the effect. If $n$ bosons are present, the probability that one more particle will ricochet in the same direction is $n + 1$ times bigger than if no other bosons are present. Talk about herd behaviour! The mere presence of other bosons doing something greatly increases the probability that one more will do the same thing.

This gregariousness turns out to have important practical applications—for instance, in the propagation of light.

## LASERS AND LIQUIDS THAT RUN UPHILL

All the processes so far considered have involved particles colliding and ricocheting in a particular direction. But that is not essential. The arguments used could apply equally well to the creation of particles— for instance, the "creation" of photons by atoms that emit light.

Photons are bosons, so the probability that an atom will emit a photon in a particular direction with a particular energy is increased by a factor of $n + 1$ if there are already $n$ photons flying in that direction with that energy. Each new photon emitted increases the chance of another photon being emitted. Once there are thousands, even millions, flying through space together, the probability of new photons being emitted is enormously enhanced.

The consequences are dramatic. Whereas a normal light source like the Sun produces a chaotic mixture of photons of all different energies, a laser generates an unstoppable tide of photons that surge through space in perfect lockstep. Lasers, however, are far from the only consequence of the gregariousness of bosons. Take liquid helium, which is composed of atoms that are bosons.

Helium-4, the second most common atom in the Universe, is one of nature's most peculiar substances.[4] It was the only element to have been discovered on the Sun before it was discovered on Earth, and it has the lowest boiling point of any liquid, –269 degrees Celsius. In fact, it is the only liquid that never freezes to become a solid, at least not at normal atmospheric pressure. All these things, however, pale into insignificance beside the behaviour of helium below about –271 degrees Celsius. Below this "lambda point," it becomes a superfluid.

Usually, a liquid resists any attempt to move one part relative to another. For instance, treacle resists when you stir it with a spoon and

---

[4]Helium-4 has four particles in its nucleus—two protons and two neutrons. It has a less common cousin, helium-3, which has the same number of protons but one fewer neutron.

water resists when you try to swim through it. Physicists call this resistance viscosity. It is really just liquid friction. But whereas we are used to friction between solids moving relative to each other—for instance, the friction between a car's tyres and the road—we are not familiar with the friction between parts of a liquid moving relative to each other. Treacle, because it resists strongly, is said to have a high viscosity, or simply to be very viscous.

Clearly, viscosity can manifest itself only when one part of a liquid moves differently from the rest. At the microscopic level of atoms, this means that it must be possible to knock some liquid atoms into states that are different from those occupied by other liquid atoms.

In a liquid at normal temperature, the atoms can be in many possible states in each of which they jiggle about at different speeds. But as the temperature falls, they become more and more sluggish and fewer and fewer states are open to them. Despite this effect, however, not all atoms will be in the same state, even at the lowest temperatures.

But things are different for a liquid of bosons such as liquid helium. Remember, if there are already $n$ bosons in a particular state, the probability of another one entering the state is $n + 1$ bigger than if there were no other particles in the state. And for liquid helium, with countless helium atoms, $n$ is a very large number indeed. Consequently, there comes a time, as liquid helium is cooled to sufficiently low temperatures, when all the helium atoms suddenly try to crowd into the same state. It's called the Bose-Einstein condensation.

With all the helium atoms in the same state, it is impossible—or at least extremely difficult—for one part of the liquid to move differently from another part. If some atoms are moving along, all the atoms have to move along together. Consequently, the liquid helium has no viscosity whatsoever. It has become a superfluid.

In superfluid liquid helium there is a kind of rigidity to the motion of the atoms. It is very hard to make the liquid do anything because you either have to get all of its atoms to do the thing together or they simply do not do the thing at all. For instance, if you put water in a bucket and spin the bucket about its axis, the water will end up spinning with the bucket. This is because the bucket drags around the water atoms—strictly speaking, the water *molecules*—that are in direct contact with the sides, and these in turn drag around the atoms farther from the sides, and so on, until the entire body of water is turning with the bucket. Clearly, for the water to get to the state in which it is spinning along with the bucket, different parts of the liquid must move relative to each other. But as just pointed out, this is very hard for a superfluid. All the atoms move together or they do not move at all. Consequently, if superfluid liquid helium is put in a bucket and the bucket is spun, it has no means open to it to attain the spin of the bucket. Instead, the superfluid helium stays stubbornly still while the bucket spins.

The cooperative motion of atoms in superfluid liquid helium leads to even more bizarre phenomena. For instance, the superfluid can flow through impossibly small holes that no other liquid can flow through. It is also the only liquid that can flow uphill.

Interestingly, helium has a rare, lightweight cousin. Helium-3 turns out to be a normal, boring liquid. The reason is that helium-3 particles are fermions. And superfluidity is a property solely of bosons.

Actually, this isn't entirely true. The microscopic world is full of surprising phenomena. And in a special case, fermions can behave like bosons!

## ELECTRIC CURRENTS THAT RUN FOREVER

The special case, when fermions behave like bosons, is that of an electric current in a metal. Because the outermost electrons of metal atoms are very loosely bound, they can break free. If a voltage is then applied between the ends of the metal by a battery, all the countless liberated electrons will surge through the material as an electric current.[5]

Electrons are, of course, fermions, which means they are anti-social. Imagine a ladder, with the rungs corresponding to ever higher energy states. Electrons would fill up the rungs two at a time from the bottom (bosons would happily crowd on the lowest rungs). The need for a separate rung for each pair of electrons means that the electrons in a metal have far more energy on average than might be naively expected.

But something really weird happens when a metal is cooled to close to absolute zero, the lowest possible temperature. Usually, each electron travels through the metal entirely independently of all other electrons. However, as the temperature falls, the metal atoms vibrate ever more sluggishly. Although they are thousands of times more massive than electrons, the attractive electrical force between an electron and a metal atom is enough to tug the atom toward it as the electron passes by.[6] The tugged atom, in turn, tugs on another electron. In this way, one electron attracts another through the intermediary of the metal atom.

This effect radically changes the nature of the current flowing through the metal. Instead of being composed of single electrons, it is composed of paired-up electrons known as Cooper pairs. But the electrons in each Cooper pair spin in an opposite manner and cancel out. Consequently, Cooper pairs are bosons!

---

[5]Why then doesn't a metal fall apart? The full explanation requires quantum theory. But, simplistically, the stripped, or conduction, electrons form a negatively charged cloud permeating the metal. It is the attraction between this cloud and the positively charged electron-stripped metal ions that glues the metal together.

[6]Strictly speaking, the atoms are positive ions, the name given to atoms that have lost electrons.

A Cooper pair is a peculiar thing. The electrons that make it up may not even be close to each other in the metal. There could easily be thousands of other electrons between one member of a Cooper pair and its partner. This is just a curious detail, however. The key thing is that Cooper pairs are bosons. And at the ultralow temperature of the superconductor all the bosons crowd into the same state. They therefore behave as a single, irresistible entity. Once they are flowing en masse, it is extremely difficult to stop them.

In a normal metal an electrical current is resisted by nonmetal, impurity atoms, which get in the way of electrons, obstructing their progress through the metal. But whereas an impurity atom can easily hinder an electron in a normal metal, it is nearly impossible for it to hinder a Cooper pair in a superconductor. This is because each Cooper pair is in lockstep with billions upon billions of others. An impurity atom can no more thwart this flow than a single soldier can stop the advance of an enemy army. Once started, the current in a superconductor will flow forever.

# PART TWO

# BIG THINGS

# 7

# THE DEATH OF SPACE AND TIME

### HOW WE DISCOVERED THAT LIGHT IS THE ROCK ON WHICH THE UNIVERSE IS FOUNDED AND TIME AND SPACE ARE SHIFTING SANDS

*When a man sits with a pretty girl for an hour, it seems like a minute. But let him sit on a hot stove for a minute—it's longer than an hour. That's relativity!*

Albert Einstein

*It's the most peculiar 100 metres anyone has ever seen. As the sprinters explode out of their starting blocks and get into their stride, it seems to the spectators in the grandstand that the runners get ever slimmer. Now, as they dash past the cheering crowd, they appear as flat as pancakes. But that's not the most peculiar thing—not by a long shot. The arms and legs of the athletes are pumping in ultraslow motion, as if they are running not through air but through molasses. Already, the crowd is beginning to slow-hand-clap. Some people are even ripping up their tickets and angrily tossing them into the air. At this pathetic rate of progress, it could take an hour for the sprinters to reach the finishing tape. Disgusted and disappointed, the spectators get up from their seats and, one by one, traipse out of the stadium.*

This scene seems totally ridiculous. But, actually, it is wrong in essentially only one detail—the speed of the sprinters. If they could run 10 million times faster, this is exactly what everyone would see. When objects fly past at ultrahigh speed, space shrinks while time slows down.[1] It's an inevitable consequence of one thing—the impossibility of ever catching up with a light beam.

Naively, you might think that the only thing that is not catch-up-able is something travelling at infinite speed. Infinity, after all, is defined as the biggest number imaginable. Whatever number you think of, infinity is bigger. So if there were something that could travel infinitely fast, it is clear you could never get abreast of it. It would represent the ultimate cosmic speed limit.

Light travels tremendously fast—300,000 kilometres per second in empty space—but this is far short of infinite speed. Nevertheless, you can never catch up with a light beam, no matter how fast you travel. In our universe, for reasons nobody completely understands, the speed of light plays the role of infinite speed. It represents the ultimate cosmic speed limit.

The first person to recognise this peculiar fact was Albert Einstein. Reputedly at the age of only 16, he asked himself: What would a beam of light look like if you could catch up with it?

Einstein could ask such a question and hope to answer it only because of a discovery made by the Scottish physicist James Clerk Maxwell. In 1868, Maxwell summarised all known electrical and magnetic phenomena—from the operation of electric motors to the behaviour of magnets—with a handful of elegant mathematical equations. The unexpected bonus of Maxwell's equations was that they predicted the existence of a hitherto unsuspected wave, a wave of electricity and magnetism.

Maxwell's wave, which propagated through space like a ripple spreading on a pond, had a very striking feature. It travelled at 300,000

---

[1]Strictly speaking, each runner will also appear to rotate, so the spectators will see some of the far side of each of them—the side facing away from the grandstand, which would normally be hidden. This peculiar effect is known as relativistic aberration, or relativistic beaming. However, it is beyond the scope of this book.

kilometres per second—the same as the speed of light in empty space. It was too much of a coincidence. Maxwell guessed—correctly—that the wave of electricity and magnetism was none other than a wave of light. Nobody, apart perhaps from the electrical pioneer Michael Faraday, had the slightest inkling that light was connected with electricity and magnetism. But there it was, written indelibly in Maxwell's equations: light was an electromagnetic wave.

Magnetism is an invisible force field that reaches out into the space surrounding a magnet. The magnetic field of a bar magnet, for instance, attracts nearby metal objects such as paperclips. Nature also boasts an electric field, an invisible force field that extends into the space around a body that is electrically charged. The electric field of a plastic comb rubbed against a nylon sweater, for instance, can pick up small scraps of paper.

Light, according to Maxwell's equations, is a wave rippling through these invisible force fields, much like a wave rippling through water. In the case of a water wave, the thing that changes as the wave passes by is the level of the water, which goes up and down, up and down. In the case of light, it is the strength of the magnetic and electric force fields, which grow and die, grow and die. (Actually, one field grows while the other dies, and vice versa, but that's not important here.)

Why go into such gory detail about what an electromagnetic wave is? The answer is because it is necessary in order to understand Einstein's question: What would a light beam look like if you could catch up with it?

Say you are driving a car on a motorway and you catch up with another car travelling at 100 kilometres per hour. What does the other car look like as you come abreast of it? Obviously, it appears stationary. If you wind down your window, you may even be able to shout to the other driver above the noise of the engine. In exactly the same way, if you could catch up with a light beam, it ought to appear stationary, like a series of ripples frozen on a pond.

However—and this is the key thing noticed by the 16-year-old Einstein—Maxwell's equations have something important to say about a frozen electromagnetic wave, one in which the electric and magnetic fields never grow or fade but remain motionless forever. No such thing exists! A stationary electromagnetic wave is an impossibility.

Einstein, with his precocious question, had put his finger on a paradox, or inconsistency, in the laws of physics. If you were able to catch up with a beam of light, you would see a stationary electromagnetic wave, which is impossible. Since seeing impossible things is, well, impossible, you can never catch up with a light beam! In other words, the thing that is uncatchable—the thing that plays the role of infinite speed in our Universe—is light.

## FOUNDATION STONES OF RELATIVITY

The uncatchability of light can be put another way. Imagine that the cosmic speed limit really is infinity (though, of course, we now know it isn't). And say for instance, a missile is fired from a fighter plane that can fly at infinite speed. Is the speed of the missile relative to someone standing on the ground infinity plus the speed of the plane? If it is, the missile's speed relative to the ground is greater than infinity. But this is impossible since infinity is the biggest number imaginable. The only thing that makes sense is that the speed of the missile is still infinitely fast. In other words, its speed does not depend on the speed of its source—the speed of the fighter plane.

It follows that in the real Universe, where the role of infinite speed is played by the speed of light, the speed of light does not depend on the motion of its source either. It's the same—300,000 kilometres per second—no matter how fast the light source is travelling.

The speed of light's lack of dependence on the motion of its source is one of the two pillars on which Einstein, in his "miraculous year" of 1905, proceeded to build a new and revolutionary picture of space and time—his "special" theory of relativity. The other one—equally important—is the principle of relativity.

In the 17th century the great Italian physicist Galileo noticed that the laws of physics are unaffected by relative motion. In other words, they appear the same, no matter how fast you are moving relative to someone else. Think of standing in a field and throwing a ball to a friend 10 metres away. Now imagine you are on a moving train instead and throwing the ball to your friend, who is standing 10 metres along the aisle. The ball in both cases loops between you on a similar trajectory. In other words, the path the ball follows takes no account of the fact that you are in a field or on a train barrelling along at, say 120 kilometres per hour.

In fact, if the windows of the train are blacked out, and the train has such brilliant suspension that it is vibration free, you will be unable to tell from the motion of the ball—or any other object inside the train, for that matter—whether or not the train is moving. For reasons nobody knows, the laws of physics are the same no matter what speed you are travelling, as long as that speed remains constant.

When Galileo made this observation, the laws he had in mind were the laws of motion that govern such things as the trajectory of cannonballs flying through the air. Einstein's audacious leap was to extend the idea to all laws of physics, including the laws of optics that govern the behaviour of light. According to his principle of relativity, all laws appear the same for observers moving with constant speed relative to each other. In a blacked-out train, in other words, you could not tell even from the way light was reflected back and forth whether or not the train was moving.

By combining the principle of relativity with the fact that the speed of light is the same irrespective of the motion of its source, it is possible to deduce another remarkable property of light. Say you are travelling towards a source of light at high speed. At what speed does the light come towards you? Well, remember there is no experiment you can do to determine whether it is you or the light source that is moving (recall the blacked-out train). So an equally valid point of view is to assume that you are stationary and the light source is moving towards you. But remember, the speed of light does not depend on the speed of its source. It always leaves the source at precisely

300,000 kilometres per second. Since you are stationary, therefore, the light must arrive at precisely 300,000 kilometres per second.

Consequently, not only is the speed of light independent of the motion of its source, it is also independent of the motion of anyone observing the light. In other words, everyone in the Universe, no matter how fast they are moving, always measures exactly the same speed of light—300,000 kilometres per second.

What Einstein set out to answer in his special theory of relativity was how, in practice, everyone can end up measuring precisely the same speed for light. It turns out there is only one way: If space and time are totally different from what everyone thinks they are.

## SHRINKING SPACE, STRETCHY TIME

Why do space and time come into things? Well, the speed of anything—light included—is the distance in space a body travels in a given interval of time. Rulers are commonly used to measure distance and clocks to measure time. Consequently, the question—how can everyone, no matter what their state of motion, measure the same speed of light?—can be put another way. What must happen to everyone's rulers and clocks so that, when they measure the distance light travels in a given time, they always get a speed of exactly 300,000 kilometres per second?

This, in a nutshell, is special relativity—a "recipe" for what must happen to space and time so that everyone in the Universe agrees on the speed of light.

Think of a spaceship firing a laser beam at a piece of space debris that happens to be flying toward it at 0.75 times the speed of light. The laser beam cannot hit the debris at 1.75 times the speed of light because that is impossible; it must hit it at exactly the speed of light. The only way this can happen is if someone observing the events and estimating the distance that the arriving light travels in a given time either underestimates the distance or overestimates the time.

In fact, as Einstein discovered, they do both. To someone watching the spaceship from outside, moving rulers shrink and moving clocks slow down. Space "contracts" and time "dilates," and they contract and dilate in exactly the manner necessary for the speed of light to come out as 300,000 kilometres per second for everyone in the Universe. It's like some huge cosmic conspiracy. The constant thing in our Universe isn't space or the flow of time but the speed of light. And everything else in the Universe has no choice but to adjust itself to maintain light in its preeminent position.

Space and time are both relative. Lengths and time intervals become significantly warped at speeds approaching the speed of light. One person's interval of space is not the same as another person's interval of space. One person's interval of time is not the same as another person's interval of time.

Time, it turns out, runs at different rates for different observers, depending on how fast they are moving relative to each other. And the discrepancy between the ticking of their clocks gets greater the speedier the motion. The faster you go, the slower you age![2] It's a truth that has been hidden from us for most of human history for the simple reason that the slowing down of time is apparent only at speeds approaching that of light, and the speed of light is so enormous that a supersonic jet, by comparison, flies at a snail's pace across the sky. If the speed of light had instead been only 30 kilometres per hour, it would not have taken a genius like Einstein to discover the truth. The effects of special relativity such as time dilation and length contraction would be glaringly obvious to the average 5-year-old.

As with time, so with space. The spatial distance between any two bodies is different for different observers, depending on how fast they are moving relative to each other. And the discrepancy between their rulers gets greater the faster the motion. "The faster you go, the slim-

---

[2]To be precise, a stationary observer sees time slow down for a moving observer by a factor $\gamma$, where $\gamma = 1/\sqrt{(1 - (v^2/c^2))}$ and $v$ and $c$ are the speed of the moving observer and the speed of light, respectively. At speeds close to $c$, $\gamma$ becomes enormous and time for a moving observer slows almost to a standstill!

mer you are," said Einstein.[3] Once again, this would be self-evident if we lived our lives travelling close to the speed of light. But living as we do in nature's slow lane, we cannot see the truth—that space and time are shifting sand, the unvarying speed of light the bedrock on which the Universe is built.

(If you think relativity is hard, take heart from the words of Einstein: "The hardest thing in the world to understand is income tax!" Ignore, however, the words of Israel's first president, Chaim Weizmann, who, after a sea voyage with the great scientist in 1921, said: "Einstein explained his theory to me every day and, on my arrival, I was fully convinced that he understood it!")

Can anything travel faster than light? Well, nothing can catch up with a beam of light. But the possibility exists that there are "sub-atomic" particles that live their lives permanently travelling faster than light. Physicists call such hypothetical particles tachyons. If tachyons exist, perhaps in the far future we can find a way to change the atoms of our bodies into tachyons and then back again. Then we too could travel faster than light.

One of the problems with tachyons, however, is that from the point of view of certain moving observers, a body travelling faster than light could appear to be travelling back in time! There is a limerick that goes like this:

> *A rocket explorer named Wright,*
> *Once travelled much faster than light.*
> *He set out one day, in a relative way,*
> *And returned on the previous night!*

                                                            Anonymous

---

[3]To be precise, a stationary observer sees the length of a moving body shrink by a factor $\gamma$, where $\gamma = 1/\sqrt{(1 - (v^2/c^2))}$ and $v$ and $c$ are the speed of the moving observer and the speed of light, respectively. At speeds close to $c$, $\gamma$ becomes enormous and a body becomes as flat as a pancake in the direction of its motion!

Time travel scares the living daylights out of physicists because it raises the possibility of paradoxes, events that lead to logical contradictions like you going back in time and killing your grandfather. If you killed your grandfather before he conceived your mother, goes the argument, how could you have been born to go back in time to kill your grandfather? Some physicists, however, think that some as-yet-undiscovered law of physics intervenes to prevent any paradoxical things from happening, and so time travel may be possible.

## THE MEANING OF RELATIVITY

But what does relativity mean in a nuts-and-bolts sense? Well, say it were possible for you to travel to the nearest star and back at 99.5 per cent of the speed of light. Since Alpha Centauri is about 4.3 light-years from Earth, those left on Earth will see you return after about 9 years, assuming a brief stopover to see the sights. From your point of view, however, the distance to Alpha Centauri will be shrunk by 10 times because of relativity. Consequently, the round-trip will take only nine-tenths of a year, or about 11 months. Say you departed on your journey on your twenty-first birthday, waved off from the spaceport by your identical twin brother. When you arrived back home, now almost 22 years old, your twin would be 30![4]

-------

[4] Actually, there is a subtle flaw in this argument. Since motion is relative, it is perfectly justifiable for your Earth-bound twin to assume that it is Earth that receded from your spacecraft at 99.5 per cent of the speed of light. However, this viewpoint leads to the opposite conclusion than before—that time slows for your twin relative to you. Clearly, time cannot run slowly for each of you, with respect to the other. The resolution of this twin paradox, as it is known, is to realise that your spaceship actually has to slow down and reverse its motion at Alpha Centauri. Because of this deceleration, the two points of view—your spaceship moving or Earth moving—are not really equivalent and interchangeable.

How would your stay-at-home twin make sense of this state of affairs? Well, he would assume that you had been living in slow motion throughout your journey. And, sure enough, if it were somehow possible for him to observe you inside your spaceship, he would see you moving as if through treacle, with all the shipboard clocks crawling around 10 times slower than normal. Your twin will correctly attribute this to the time dilation of relativity. But to you all the clocks and everything else on board will appear to be moving at perfectly normal speed. This is the magic of relativity.

Of course, the more rapidly you travelled to Alpha Centauri and back, the greater the discrepancy between your age and your twin's. Travel fast enough and far enough across the Universe and you will return to find that your twin is long dead and buried. Even faster and you will find that Earth itself has dried up and died. In fact, if you travelled within a whisker of the speed of light, time would go so slowly for you that you could watch the entire future history of the Universe flash past you like a movie in fast-forward. "The possibility of visiting the future is quite awesome to anyone who learns about it for the first time," says Russian physicist Igor Novikov.

We do not yet have the ability to travel to the nearest star and back at close to the speed of light (or even 0.01 per cent of the speed of light). Nevertheless, time dilation is detectable—just—in the everyday world. Experiments have been carried out in which super-accurate atomic clocks are synchronised and separated, one being flown around the world on an airplane while the other stays at home. When the clocks are reunited, the experimenters find that the around-the-world clock has registered the passage of marginally less time than its stay-at-home counterpart. The shorter time measured by the moving clock is precisely what is predicted by Einstein.

The slowing of time affects astronauts too. As Novikov points out in his excellent book, *The River of Time:* "When the crew of the Soviet Salyut space station returned to Earth in 1988 after orbiting for a year at 8 kilometres a second, they stepped into the future by one hundredth of a second."

The time dilation effect is minuscule because airplanes and space-craft travel at only a tiny fraction of the speed of light. However, it is far greater for cosmic-ray muons, subatomic particles created when cosmic rays—superfast atomic nuclei from space—slam into air molecules at the top of Earth's atmosphere.

The key thing to know about muons is that they have tragically short lives and, on average, disintegrate, or decay after a mere 1.5 millionths of a second. Since they streak down through the atmosphere at more than 99.92 per cent of the speed of light, this means that they should travel barely 0.5 kilometres before self-destructing. This is not far at all when it is realised that cosmic-ray muons are created about 12.5 kilometres up in the air. Essentially none, therefore, should reach the ground.

Contrary to all expectations, however, every square metre of Earth's surface is struck by several hundred cosmic-ray muons every second. Somehow, these tiny particles manage to travel 25 times farther than they have any right to. And it is all because of relativity.

The time experienced by a speeding muon is not the same as the time experienced by someone on Earth's surface. Think of a muon as having an internal alarm clock that tells it when to decay. At 99.92 per cent of the speed of light, the clock slows down by a factor of about 25, at least to an observer on the ground. Consequently, cosmic-ray muons live 25 times longer than they would if stationary—time enough to travel all the way to the ground before they disintegrate. Cosmic-ray muons on the ground owe their very existence to time dilation.

What does the world look like from a muon's point of view? Or come to think of it, from the point of view of the space-faring twin or the atomic clock flown round the world? Well, from the point of view of all of these, time flows perfectly normally. Each, after all, is stationary with respect to itself. Take the muon. It still decays after 1.5 millionths of a second. From its point of view, however, it is standing still and it's Earth's surface that is approaching at 99.92 per cent of the speed of light. It therefore sees the distance it has to travel shrink by a factor of 25, enabling it to reach the ground even in its ultrashort lifetime.

The great cosmic conspiracy between time and space works whatever way you look at it.

## WHY RELATIVITY HAD TO BE

The behaviour of space and time at speeds approaching that of light is indeed bizarre. However, it need not have been a surprise to anyone. Although our everyday experience in nature's slow lane has taught us that one person's interval of time is another person's interval of time and that one person's interval of space is another person's interval of space, our belief in both of these things is in fact based on a very rickety assumption.

Take time. You can spend a lifetime trying futilely to define it. Einstein, however, realised that the only useful definition is a practical one. We measure the passage of time with watches and clocks. Einstein therefore said: "Time is what a clock measures." (Sometimes, it takes a genius to state the obvious!)

If everyone is going to measure the same interval of time between two events, this is equivalent to saying that their clocks run at the same rate. But as everyone knows, this never quite happens. Your alarm clock may run a little slow, your watch a little fast. We overcome these problems by, now and then, synchronising them. For instance, we ask someone the right time and, when they tell us, we correct our watch accordingly. Or we listen for the time signal "pips" on the BBC. But in using the pips, we make a hidden assumption. The assumption is that it takes no time at all for the radio announcement to travel to our radio. Consequently, when we hear the radio announcer say it is 6 a.m., it is 6 a.m.

A signal that takes no time at all travels infinitely fast. The two statements are entirely equivalent. But as we know, there is nothing in our Universe that can travel with infinite speed. On the other hand, the speed of radio waves—a form of light invisible to the naked eye—

is so huge compared to all human distances that we notice no delay in their travel to us from the transmitter. Our assumption that the radio waves travel infinitely fast, although false, is not a bad one in the circumstances. But what happens if the distance from the transmitter is very large indeed? Say the transmitter is on Mars.

When Mars is at its closest, the signal takes 5 minutes to fly across space to Earth. If, when we hear the announcer on Mars say it is 6 a.m., we set our clock to 6 a.m., we will be setting it to the wrong time. The way around this is obviously to take into account the 5-minute time delay and, when we hear 6 a.m., set our clock to 6:05.

Everything, of course, hinges on knowing the time it takes for the signal to travel from Earth to Mars. In practice this can be done by bouncing a radio signal from Earth off Mars and picking up the return signal. If it takes 10 minutes for the round-trip, then it must take 5 minutes to travel from the spaceship to Earth.

The lack of an infinitely fast means of sending signals is not, therefore, a problem in itself for synchronising everyone's clocks. It can still be done by bouncing light signals back and forth and taking into account the time delays. The trouble is that this works perfectly only if everyone is stationary with respect to everyone else. In reality, everyone in the Universe is moving with respect to everyone else. And the minute you start bouncing light signals between observers who are moving, the peculiar constancy of the speed of light starts to wreak havoc with common sense.

Say there is a spaceship travelling between Earth and Mars and say it is moving so fast that, by comparison, Earth and Mars appear stationary. Imagine that, as before, you send a radio signal to Mars, which bounces off the planet and which you then pick up back on Earth. The round-trip takes 10 minutes, so, as before, you deduce that the signal arrived at Mars after only 5 minutes. Once again, if you pick up a time signal from Mars, saying it is 6 a.m., you will deduce from the time delay that it is really 6:05.

Now consider the spaceship. Assume that at the instant you send your radio signal to Mars, it sets off at its full speed to Mars. At what time does an observer on the spaceship see the radio signal arrive at Mars?

Well, from the observer's point of view, Mars is approaching, so the radio signal has a shorter distance to travel. But the speed of the signal is the same for you and for the observer on the spaceship. After all, that's the central peculiarity of light—it has exactly the same speed for everyone.

Speed, remember, is simply the distance something travels in a given time. So if the observer on the spaceship sees the radio signal travel a shorter distance and still measures the same speed, the observer must measure a shorter time too. In other words, the observer deduces that the radio signal arrives at Mars earlier than you deduce it does. To the observer, clocks on Mars tick more slowly. If the observer picks up a time signal from Mars, saying it is 6 a.m., the observer will correct it using a shorter time delay and conclude it is, say 6:03, not the 6:05 you conclude.

The upshot is that two observers who are moving relative to each other never assign the same time to a distant event. Their clocks always run at different speeds. What is more, this difference is absolutely fundamental—no amount of ingenuity in synchronising clocks can ever get around it.

## SHADOWS OF SPACE-TIME

The slowing of time and the shrinking of space is the price that must be paid so that everyone in the Universe, no matter what their state of motion, measures the same speed of light. But this is only the beginning.

Say there are two stars and a space-suited figure is floating in the blackness midway between them. Imagine that the two stars explode and the floating figure sees them detonate simultaneously—two blinding flashes of light on either side of him. Now picture a spaceship travelling at enormous speed along the line joining the two stars. The spaceship passes by the space-suited figure just as he sees the two stars explode. What does the pilot of the spaceship see?

Well, since the ship is moving towards one star and away from the other, the light from the star it is approaching will arrive before the light from the star it is receding from. The two explosions will therefore not appear simultaneously. Consequently, even the concept of simultaneity is a casualty of the constancy of the speed of light. Events that one observer sees as simultaneous are not simultaneous to another observer moving with respect to the first.

The key thing here is that the exploding stars are separated by an interval of space. Events that one person sees separated by only space, another person sees separated by space and time—and vice versa. Events one person sees separated only by time, another person sees separated by time and space.

The price of everyone measuring the same speed of light is therefore not only that the time of someone moving past you at high speed slows down while their space shrinks but that some of their space appears to you as time and some of their time appears to you as space. One person's interval of space is another person's interval of space and time. And one person's interval of time is another person's interval of time and space. The fact that space and time are interchangeable in this way tells us something remarkable and unexpected about space and time. Fundamentally, they are same thing—or at least different sides of the same coin.

The person who first saw this—more clearly even than Einstein himself—was Einstein's former mathematics professor Hermann Minkowski, a man famous for calling his student a "lazy dog" who would never amount to anything. (To his eternal credit, he later ate his words.) "From now on," said Minkowski, "space of itself and time of itself will sink into mere shadows and only a kind of union between them will survive."

Minkowski christened this peculiar union of space and time "space-time." Its existence would be blatantly obvious to us if we lived our lives travelling at close to the speed of light. Living as we do in nature's ultraslow lane, however, we never experience the seamless entity. All we glimpse instead are its space and time facets.

As Minkowski put it, space and time are like shadows of space-time. Think of a stick suspended from the ceiling of a room so that it can spin around its middle and point in any direction like a compass needle. A bright light casts a shadow of the stick on one wall while a second bright light casts a shadow of the object on an adjacent wall. We could, if we wanted, call the size of the stick's shadow on one wall its "length" and the size of its shadow on the other wall its width. What then happens as the stick swings around?

Clearly, the size of the shadow on each wall changes. As the length gets smaller, the width gets bigger, and vice versa. In fact, the length appears to change into the width and the width into the length—just as if they are aspects of the same thing.

Of course, they are aspects of the same thing. The length and width are not fundamental at all. They are simply artifacts of the direction from which we choose to observe the stick. The fundamental thing is the stick itself, which we can see simply by ignoring the shadows on the wall and walking up to it at the centre of the room.

Well, space and time are much like the length and width of the stick. They are not fundamental at all but are artifacts of our viewpoint—specifically, how fast we are travelling. But though the fundamental thing is space-time, this is apparent only from a viewpoint travelling close to the speed of light, which of course is why it is not obvious to any of us in our daily lives.

Of course, the stick-and-shadow analogy, like all analogies, is helpful only up to a point. Whereas the length and width of the stick are entirely equivalent, this is not quite true of the space and the time facets of space-time. Though you can move in any direction you like in space, as everyone knows you can only move in one direction in time.

The fact that space-time is solid reality and space and time the mere shadows raises a general point. Like shipwrecked mariners clinging to rocks in a wild sea, to make sense of the world we search desperately for things that are unchanging. We identify things like distance and time and mass. But later, we discover that the things we identified as unchanging are unchanging only from our limited viewpoint. When we widen our perspective on the world we discover that other things we never suspected are the invariant things. So it is with space and time. When we see the world from a high-speed vantage point, we see neither space nor time but the seamless entity of space-time.

Actually, we should long ago have guessed that space and time are inextricably entwined. Think of the Moon. What is it like now, at this instant? The answer is that we can never know. All we can ever know is what it was like $1^1/_4$ seconds ago, which is the time it takes light from the Moon to fly across the 400,000 kilometres to Earth. Now think of the Sun. We cannot know what it is like either, only what it was like $8^1/_2$ minutes ago. And for the nearest star system, Alpha Centauri, it is even worse. We have to make do with a picture that by the time we see it is already 4.3 years out of date.

The point is that, although we think of the Universe we see through our telescopes as existing now, this is a mistaken view. We can never know what the Universe is like at this instant. The farther across space we look, the farther back in time we see. If we look far enough across space we can actually see close to the Big Bang itself, 13.7 billion years back in time. Space and time are inextricably bound together. The Universe we see "out there" is not a thing that extends in space but a thing that extends in space-time.

The reason we have been hoodwinked into thinking of space and time as separate things is that light takes so little time to travel human distances that we rarely notice the delay. When you are talking with someone, you see them as they were a billionth of a second earlier. But this interval is unnoticeable because it is 10 million times shorter than any event that can be perceived by the human brain. It is no wonder that we have come to believe that everything we perceive around us exists now. But "now" is a fictitious concept, which becomes obvious as soon as we contemplate the wider universe, where distances are so great that it takes light billions of years to span them.

The space-time of the Universe can be thought of as a vast map. All events—from the creation of the Universe in the Big Bang to your birth at a particular time and place on Earth—are laid out on it, each with its unique space-time location. The map picture is appropriate because time, as the flip side of space, can be thought of as an additional spatial dimension. But the map picture poses a problem. If everything is laid out—preordained almost—there is no room for the concepts of past, present, and future. As Einstein remarked: "For us physicists, the distinction between past, present, and future is only an illusion."

It is a pretty compelling illusion, though. Nevertheless, the fact remains that the concepts of past, present, and future do not figure at all in special relativity, one of our most fundamental descriptions of reality. Nature appears not to need them. Why we do is one of the great unsolved mysteries.

## $E = mc^2$ AND ALL THAT

The special theory of relativity does more than profoundly change our ideas of space and time. It changes our ideas about a host of other things too. The reason is that all the basic quantities of physics are founded on space and time. If, as relativity tells us, space and time are malleable, blurring one into the other as the speed of light is approached, then so too are the other entities—momentum and energy, electric fields and magnetic fields. Like space and time, which merge into the seamless medium of space-time, they too are inextricably tied together in the interests of keeping the speed of light constant.

Take electricity and magnetism. It turns out that, just as one person's space is another person's time, one person's magnetic field is another person's electric field. Electric and magnetic fields are crucial to both generators that make electrical currents and motors that turn electric currents into motion. "The rotating armatures of every generator and every motor in this age of electricity are steadily proclaiming the truth of the relativity theory to all who have ears to hear," wrote the physicist Leigh Page in the 1940s. Because we live in a slow-motion world, we are hoodwinked into believing that electric and magnetic fields have separate existences. But just like space and time, they are merely different faces of the same coin. In reality there is only a seamless entity: the electromagnetic field.

Two other quantities that turn out to be different faces of the same coin are energy and momentum.[5] And in this unlikely connection is hidden perhaps the greatest surprise of relativity—that mass is a form of energy. The discovery is encapsulated in the most famous, and least understood, formula in all of science: $E = mc^2$.

---

[5]The momentum of a body is a measure of how much effort is required to stop it. For instance, an oil tanker, even though it may be moving only a few kilometres an hour, is far harder to stop than a Formula 1 racing car going 200 kilometres per hour. We say the oil tanker has more momentum.

# 8

# $E = mc^2$ AND THE WEIGHT OF SUNSHINE

## HOW WE DISCOVERED THAT ORDINARY MATTER CONTAINS A MILLION TIMES THE DESTRUCTIVE POWER OF DYNAMITE

*Photons have mass?!? I didn't even know they were Catholic.*

Woody Allen

*It's the biggest set of bathroom scales imaginable. And, oh, yes, it's heat resistant too. It's so big in fact that it can weigh a whole star. And today it's weighing the nearest star of all: our Sun. The digital display has just come to rest and it's registering $2 \times 10^{27}$ tons. That's 2 followed by 27 zeroes—2,000 million million million million tons. But wait a minute, something's wrong. The scales are superaccurate. That's another remarkable thing about them, in addition to their size and heat resistance! But every second, when the display is refreshed, it reads 4 million tons less than it did the previous second. What's going on? Surely the Sun isn't really getting lighter—by the weight of a good-sized supertanker—every single second?*

Ah, but it is! The Sun is losing heat-energy, radiating it into space as sunlight. And energy actually *weighs* something.[1] So the more sun-

---

[1] I am using the word weight here the way it is used in everyday life as synonymous with mass. Strictly speaking, weight is equivalent to the force of gravity.

light the Sun gives out, the lighter it gets. Mind you, the Sun is *big* and we're only talking about it losing about a *10-million-million-millionth* of a per cent of its mass per second. That's hardly more than 0.1 per cent of its mass since its birth.

The fact that energy does indeed weigh something can be seen vividly from the behaviour of a comet. The tail of a comet always points away from the Sun just like a windsock points away from the gathering storm.[2] What do the two have in common? Both are being pushed by a powerful wind. In the case of the windsock, it's a wind of air; in the case of the comet tail, a wind of light streaming outward from the Sun.

The windsock is being hit by air molecules in their countless trillions. It is this relentless bombardment that is pushing the fabric and causing it to billow outward. The story is pretty much the same out in deep space. The comet tail is being battered by countless tiny particles of light. It is the machine-gun bombardment of these photons that is causing the glowing cometary gases to billow across tens of millions of kilometres of empty space.[3]

But there is an important difference between the windsock being struck by air molecules and the comet's tail being hit by photons. The air molecules are solid grains of matter. They thud into the material of the windsock like tiny bullets, and this is why the windsock recoils. But photons are not solid matter. They actually have no mass. How then can they be having a similar effect to air molecules, which do?

---

[2] A comet is a giant interplanetary snowball. Billions of such bodies are believed to orbit in the deep freeze beyond the outermost planet. Occasionally, one is nudged by the gravity of a passing star and falls toward the Sun. As it heats up, its surface cracks, and buckles, and boils off into the vacuum to form a long, glowing tail of gas.

[3] Actually, the tail of a comet is pushed by a combination of the light from the Sun and the solar wind, the million-mile-an-hour hurricane of subatomic particles—mostly hydrogen nuclei—that streams out from the Sun.

Well, one thing photons certainly do have is energy. Think of the heat that sunlight deposits on your skin when you sunbathe on a summer's day. The inescapable conclusion is that the energy must actually *weigh* something.[4]

This turns out to be a direct consequence of the uncatchability of light. Because the speed of light is terminally out of reach, no material body can ever be accelerated to the speed of light, no matter how hard it is pushed. The speed of light, recall, plays the role of infinite speed in our Universe. Just as it would take an infinite amount of energy to accelerate a body to infinite speed, it would take an infinite amount of energy to push one to light speed. In other words, the reason that getting to the speed of light is impossible is because it would take more energy than is contained in the Universe.

What would happen, however, if you were to push a mass closer and closer to the speed of light? Well, since the ultimate speed is unattainable, the body would have to become harder and harder to push as you get it closer and closer to the ultimate speed.

Being hard to push is the same as having a big mass. In fact, the mass of a body is defined by precisely this property—how hard it is to push it. A loaded refrigerator which is difficult to budge, is said to have a large mass, whereas a toaster, which is easy to budge, is said to have a small mass. It follows therefore that, if a body gets harder to push as it approaches the speed of light, it must get more massive. In fact, if a material body was ever to attain the speed of light itself, it would acquire an infinite mass, which is just another way of saying its acceleration would take an infinite amount of energy. Whatever way you look at it, it's an impossibility.

Now, it is a fundamental law of nature that energy can neither be created or destroyed, only transformed from one guise into another.

---

[4]Strictly speaking, the thing photons possess is momentum. In other words, it takes an effort to stop them. This effort is provided by the comet's tail, which recoils as a result.

For instance, electrical energy changes into light energy in a lightbulb; sound energy changes into the energy of motion of a vibrating diaphragm in a microphone. What, then, happens to the energy put into pushing a body that is moving close to the speed of light? Hardly any of the energy can go into increasing the body's speed since a body moving at close to the speed of light is already travelling within a whisker of the ultimate speed limit.

The only thing that increases as the body is pushed harder and harder is its mass. This, then, must be where all the energy goes. But, recall, energy can only be changed from one form into another. The inescapable conclusion, discovered by Einstein, is therefore that mass itself is a *form of energy*. The formula for the energy locked up in a chunk of matter of mass, *m*, is given by perhaps the most famous equation in all of science: $E = mc^2$, where *c* is the scientists' shorthand for the speed of light.

The connection between energy and mass is perhaps the most remarkable of all the consequences of Einstein's special theory of relativity. And like the connection between space and time, it is a two-way thing. Not only is mass a form of energy, but energy has an effective mass. Put crudely, *energy weighs something*.

Sound energy, light energy, electrical energy—any form of energy you can think of—they all weigh something. When you warm up a pot of coffee, you add heat-energy to it. But heat-energy weighs something. Consequently, a cup of coffee weighs slightly more when hot than when cold. The operative word here is slightly. The difference in weight is far too small to measure. In fact, it is far from obvious that energy has a weight, which is of course why it took the genius of Einstein to first notice it. Nevertheless, one form of energy at least—the energy of sunlight—does reveal its mass when it interacts with a comet.

Light can push the tail of a comet because light energy weighs something. Photons have an effective mass by virtue of their energy.

Another familiar form of energy is energy of motion. If you step into the path of a speeding cyclist, you will be left in no doubt that such a thing exists. Energy of motion, like all other forms of energy, weighs something. So you weigh marginally more when you are running than when you are walking.

It is energy of motion that explains why the photons of sunlight can push a comet tail. An explanation is needed because they actually have no *intrinsic mass*. If they did, after all, they would be unable to travel at the speed of light, a speed that is forbidden to all bodies with mass. What light has instead is an effective mass—a mass by virtue of the fact that it has energy of motion.

The existence of energy of motion also explains why a cup of coffee is heavier when hot than when cold. Heat is microscopic motion. The atoms in a liquid or solid jiggle about, while the atoms in a gas fly hither and thither. Because the atoms in a cup of hot coffee are jiggling faster than the atoms in a stone-cold cup, they possess more energy of motion. Consequently, the coffee weighs more.

## RABBITS OUT OF HATS

So much for energy having an equivalent mass, or weighing something. The fact that mass is a form of energy also has profound implications. Since one form of energy can be converted into another, mass-energy can be transformed into other forms of energy and, conversely, other forms of energy can be changed into mass-energy.

Take the latter process. If mass-energy can be made out of other forms of energy, it follows that mass can pop into existence where formerly no mass existed. This is exactly what happens in giant particle accelerators, or atom smashers. At CERN, the European centre for particle physics near Geneva, for instance, subatomic particles— the building blocks of atoms—are whirled around a subterranean racetrack and slammed together at speeds approaching that of light. In the violent smash-up, the tremendous energy of motion of the particles is converted into mass-energy—the mass of new particles that physicists wish to study. At the collision point, these particles appear apparently out of nothing, like rabbits out of a hat.

This phenomenon is an instance of one type of energy changing into mass-energy. But what about mass-energy changing into another type of energy? Does that happen? Yes, all the time.

## A MILLION TIMES THE DESTRUCTIVE POWER OF DYNAMITE

Think of a piece of burning coal. Because the heat it gives out weighs something, the coal gradually loses mass. So if it were possible to collect and weigh all the products of burning—the ash, the gases given off, and so on—they would turn out to weigh less than the original lump of coal.

The amount of mass-energy turned into heat-energy when coal burns is so small as to be essentially unmeasurable. Nevertheless, there is a place in nature where a significant mass is converted into other forms of energy. It was identified by the English physicist Francis Aston in 1919 while he was "weighing" atoms.

Recall that each of the 92 naturally occurring atoms contains a nucleus made from two distinct subatomic particles—the proton and neutron.[5] Since the masses of these two nucleons are essentially the same, the nucleus, at least as far as its weight is concerned, can be thought of as being made from a single building block. Think of it as a Lego brick. Hydrogen, the lightest nucleus, is therefore made from one Lego brick; uranium, the heaviest, is made from 238 Lego bricks.

Now, there had been a suspicion since the beginning of the 19th century that perhaps the Universe had started out with only one kind of atom—the simplest, hydrogen. Since that time, all the other atoms have somehow been built up from hydrogen, by the process of sticking together hydrogen Lego bricks. The evidence for the idea, which

---

[5]Except, of course, the most common isotope of hydrogen, the nucleus of which consists simply of one proton and no neutrons.

had been proposed by a London physician named William Prout in 1815, was that an atom like lithium appeared to weigh exactly six times as much as hydrogen, an atom like carbon exactly 12 times as much, and so on.

However, when Aston compared the masses of different kinds of atoms more precisely with an instrument he invented called a mass spectrograph, he discovered something different. Lithium in fact weighed a shade less than six hydrogen atoms; carbon weighed a shade less than 12 hydrogen atoms. The biggest discrepancy was helium, the second lightest atom. Since a helium nucleus was assembled from four Lego bricks, by rights it should weigh four times as much as a hydrogen atom. Instead, it weighed 0.8 per cent less than four hydrogen atoms. It was like putting four 1-kilogram bags of sugar on a set of scales and finding that they weighed almost 1 per cent less than 4 kilograms!

If all atoms had indeed been assembled out of hydrogen atom Lego bricks, as Prout strongly suspected, Aston's discovery revealed something remarkable about atom building. During the process, a significant amount of mass-energy went AWOL.

Mass-energy, like all forms of energy, cannot be destroyed. It can only be changed from one form into another, ultimately the lowest form of energy—heat-energy. Consequently, if 1 kilogram of hydrogen was converted into 1 kilogram of helium, 8 grams of mass-energy would be converted into heat-energy. Amazingly, this is a million times more energy than would be liberated by burning 1 kilogram of coal!

This factor of a million did not go unnoticed by astronomers. For millennia, people had wondered what kept the Sun burning. In the 5th century BC, the Greek philosopher Anaxagoras had speculated that the Sun was "a red-hot ball of iron not much bigger than Greece." Later, in the 19th century, the age of coal, physicists had naturally wondered whether the Sun was a giant lump of coal. It would have to be the mother of all lumps of coal! They found, however, that if it was a lump of coal, it would burn out in about 5,000 years. The trouble is that the evidence from geology and biology is that Earth—and by implication the Sun—is at least a million times older. The inescapable conclusion is that the Sun is drawing on an energy source a million times more concentrated than coal.

The man who put two and two together was English astronomer Arthur Eddington. The Sun, he guessed, was powered by atomic, or nuclear, energy. Deep in its interior it was sticking together the atoms of the lightest substance, hydrogen, to make atoms of the second lightest, helium. In the process, mass-energy was being turned into heat and light energy. To maintain the Sun's prodigious output, 4 million tons of mass—the equivalent of a million elephants—was being destroyed every second. Here, at last, was the ultimate source of sunlight.

This discussion conveniently skirts over the matter of why making a heavy atom out of a light atom converts so much mass-energy into other forms of energy. A digression may help.

Imagine you are walking past a house and a slate falls from the roof and hits you on the head. Energy is released in this process. For instance, there is the whack as the slate hits your head—sound energy. Maybe it even knocks you over. Then there is heat energy. If you could measure the temperature of the slate and your head very accurately, you would find they were slightly warmer than before.

Where did all this energy come from? The answer is from gravity. Gravity is a force of attraction between any two massive bodies. In this case, the gravity between Earth and the slate pulls them closer together.

Now, what would happen if gravity was twice as strong as it is? Clearly, the slate would be pulled towards Earth faster. It would make a bigger noise when it hit, create more heat, and so on. In short, more energy would be released. What if gravity were 10 times stronger? Well, even more energy would be unleashed. Now, what if gravity was 10,000 trillion trillion trillion times stronger? Obviously, a mind-bogglingly huge amount of energy would be released by the crashing slate (and the combination of Earth and slate would be lighter, like the helium atom).

But isn't this just fantasy? Surely, there is no force that is 10 trillion trillion trillion times stronger than gravity? Well, there is, and it is operating in each and every one of us at this very moment! It is called the nuclear force, and it is the glue that holds together the nuclei of atoms.

Imagine what would happen if you took the nuclei of two light atoms and let them fall together under the nuclear force rather like the slate and Earth falling together under gravity. The collision would be tremendously violent and an enormous amount of energy would be liberated—a million times more energy than would be released by burning the same weight of coal.

Atom building is not only the source of the Sun's energy. It is also the source of the energy of the hydrogen bomb. For that's all H-bombs do—slam together hydrogen nuclei (normally, a heavy cousin of hydrogen, but that's another story) to make nuclei of helium. The helium nuclei are lighter than the combined weight of their hydrogen building blocks, and the missing mass reappears as the tremendous heat energy of the nuclear fireball. The destructive power of a 1-megaton hydrogen bomb—about 50 times greater than the one that devastated Hiroshima—comes from the destruction of little more than a kilogram of mass. "If only I had known, I should have become a watchmaker!" said Einstein, reflecting on his role in the development of the nuclear bomb.

## TOTAL CONVERSION OF MASS INTO ENERGY

Even though Einstein downgraded mass, showing that it was merely one among countless other forms of energy, it is special in one way: It is the most concentrated form of energy known. In fact, the equation $E = mc^2$ encapsulates this fact. The physicists' symbol for the speed of light, $c$, is a big number—300 million metres per second. Squaring it—multiplying it by itself—creates an even bigger number. Applying the formula to 1 kilogram of matter shows that it contains $9 \times 10^{16}$ joules of energy—enough to lift the entire population of the world into space!

Of course, to get this kind of energy out of a kilogram of matter, it would be necessary to convert it entirely into another form of energy—that is, to destroy all of its mass. The nuclear processes in the Sun and a hydrogen bomb liberate barely 1 per cent of the energy locked up in matter. However, it turns out that nature can do far better than this.

Black holes are regions of space where gravity is so strong that light itself cannot escape—hence their blackness. They are the remnant left behind when a massive star dies, shrinking catastrophically in size until they literally wink out of existence. As matter swirls down into a black hole, like water down a plug hole, it rubs against itself, heating itself to incandescence. Energy is unleashed as both light and heat. In the special case when a black hole is spinning at its maximum possible rate, the liberated energy is equivalent to 43 per cent of the mass of the matter swirling in. This means that, pound for pound, the in-fall of matter onto a black hole is 43 times more efficient at generating energy than the nuclear processes powering the Sun or an H-bomb.

And this isn't just theory. The Universe contains objects called quasars, the superbright cores of newborn galaxies. Even our own Milky Way galaxy may have had a quasar in its heart in its wayward youth 10 billion years ago. The puzzling thing about quasars is that they often pump out the light energy of 100 normal galaxies—that's 10 million million suns—and from a tiny region smaller than our solar system. All that energy cannot be coming from stars; it would be impossible to squeeze 10 million million suns into such a small volume of space. It can only come from a giant black hole sucking in matter. Astronomers, therefore, firmly believe that quasars contain "supermassive" black holes—up to 3 billion times the mass of the Sun—that are steadily gobbling whole stars. But even black holes can convert barely half of the mass of matter into other forms of energy.

Is there a process that can convert all of the mass into energy? The answer is *yes*. Matter actually comes in two types—matter and anti-matter. It is not necessary to know anything about antimatter other than the fact that, when matter and antimatter meet, the two destroy, or annihilate each other, with 100 per cent of their mass-energy flashing instantly into other forms of energy.

Now, our Universe, for a reason nobody knows, appears to be made almost entirely of matter. This is a deep puzzle because, when tiny amounts of antimatter are made in the laboratory, their birth is always accompanied by an equal amount of matter. Because there is essentially no antimatter in the Universe, if we want any we have to make it. It's difficult. Not only do you have to put in a lot of energy to make it—as much energy as you are likely to get out!—but it tends to annihilate as soon as it meets ordinary matter, so it's difficult to accumulate a lot of it. So far scientists have managed to collect less than a billionth of a gram.

Nevertheless, if the problem of making antimatter in quantities could be cracked, we would have at our disposal the most powerful energy source imaginable. The problem with all spacecraft is that they have to take their fuel along with them. But that fuel weighs a lot. So fuel is needed to lift the fuel into space. The *Saturn V* rocket, for instance, weighed 3,000 tons and all that weight—mostly fuel—was needed to take two men to the surface of the Moon and return them safely to Earth. Antimatter offers a way out. A spacecraft would require hardly any antimatter to fuel it because antimatter contains such a tremendous amount of energy pound for pound. If we ever manage to travel to the stars, we will have to squeeze every last drop of energy out of matter. Just as in *Star Trek*, we will have to build starships powered by antimatter.

# 9

# THE FORCE OF GRAVITY DOES NOT EXIST

HOW WE DISCOVERED THE TRUTH ABOUT GRAVITY AND CAME FACE TO FACE
WITH BLACK HOLES, WORMHOLES, AND TIME MACHINES

*The breakthrough came suddenly one day. I was sitting on a chair in my patent office in Bern. Suddenly the thought struck me: If a man falls freely, he does not feel his own weight. I was taken aback. This simple thought experiment made a deep impression on me. This led me to the theory of gravity.*

Albert Einstein

They are 20-year-old twin sisters. They work in the same skyscraper in Manhattan. One is an assistant in a boutique at street level, the other a waitress at the High Roost restaurant on the 52nd floor. It's 8:30 a.m. They come through the revolving doors into the foyer and go their separate ways. One heads across the marble expanse to the ground-floor shopping mall; the other sprints into the mouth of the high-speed elevator just before the doors swish shut.

The hands of the clock above the elevator spin around. Now it's 5:30 p.m. On the ground floor the shop-assistant twin stares up at the big red indicator light as it counts down the floors. With a "ding," the doors burst open and out comes her waitress sister . . . an 85-year-old bent figure clutching a silver zimmer frame!

If you think this scenario is pure fantasy, think again. It's an exaggeration, granted, but it's an exaggeration of the truth. You really do age more slowly on the ground floor of a building than on the top floor. It's an effect of Einstein's "general" theory of relativity, the framework he came up with in 1915 to fix the shortcomings of his special theory of relativity.

The problem with the special theory of relativity is that, well, it is *special*. It relates what one person sees when looking at another person moving at constant speed relative to them, revealing that the moving person appears to shrink in the direction of their motion while their time slows down, effects that become ever more marked as they approach the speed of light. But motion at constant speed is of a very special kind. Bodies in general change their speed with time—for instance, a car accelerates away from traffic lights or NASA's space shuttle slows when it reenters Earth's atmosphere.

The question Einstein therefore set out to answer after he published his special theory of relativity in 1905 was: What does one person see when looking at another person accelerating relative to them? The answer, which took him more than a decade to obtain, was contained in the "general" theory of relativity, arguably the greatest contribution to science by a single human mind.

When Einstein embarked on his quest, one problem in particular worried him: what to do about Newton's law of gravity. Although it had stood unchallenged for almost 250 years, it was clear to Einstein that it was fundamentally incompatible with the special theory of relativity. According to Newton, every massive body tugs on every other massive body with an attractive force called gravity. For instance, there is a gravitational pull between Earth and each and every one of us; it keeps our feet glued firmly to the ground. There is a gravitational pull between the Sun and Earth, which keeps Earth trapped in orbit around the Sun. Einstein did not object to this idea. His difficulty was with the speed of gravity.

Newton assumed that the force of gravity acts instantaneously— that is, the Sun's gravity reaches out across space to Earth and Earth feels the tug of that gravity without any delay. Consequently, if the Sun were to vanish at this very moment—an unlikely scenario!—Earth would notice the absence of the Sun's gravity instantly and promptly fly off into interstellar space.

An influence that can cross the gulf between the Sun and Earth in no time at all must travel infinitely fast—instantaneous travel and infinite speed are completely equivalent. However, as Einstein discovered, nothing—and that necessarily includes gravity—can travel faster than light. Since light takes just over eight minutes to travel between the Sun and Earth, it follows that, if the Sun were to vanish suddenly, Earth would continue merrily in its orbit for at least eight and a bit minutes before spinning off to the stars.

Newton's tacit assumption that gravity reaches out across space at infinite speed is not the only serious flaw in his theory of gravity. He also assumed that the force of gravity is generated by mass. Einstein, however, discovered that all forms of energy have an effective mass, or weigh something. Consequently, all forms of energy—not just mass-energy—must be sources of gravity.

The challenge facing Einstein was, therefore, to incorporate the ideas of the special theory of relativity into a new theory of gravity and, at the same time, to generalise the special theory of relativity to describe what the world looked like to an accelerated person. It was as he contemplated these gargantuan challenges that a lightbulb lit up in Einstein's head. He realised, to his surprise and delight, that the two tasks were one and the same.

## THE ODD THING ABOUT GRAVITY

To understand the connection it is necessary to appreciate a peculiar property of gravity. All bodies, irrespective of their mass, fall at the same rate. A peanut, for instance, picks up speed just as quickly as a person. This behaviour was first noticed by the 17th-century Italian scientist Galileo. In fact, Galileo is reputed to have demonstrated the effect by taking a light object and a heavy object and dropping them together from the top of the Leaning Tower of Pisa. Reportedly, they hit the ground at the same time.

On Earth the effect is obscured because objects with a large surface area are preferentially slowed by their passage through the air. Nevertheless, Galileo's experiment can be carried out in a place where there is no air resistance to mess things up—the Moon. In 1972, *Apollo 15* commander Dave Scott dropped a hammer and a feather together. Sure enough, they hit the lunar soil at exactly the same time.

What is peculiar about this phenomenon is that, usually, the way in which a body moves in response to a force depends on its mass. Imagine a wooden stool and a loaded refrigerator standing on an ice rink, where there is no friction to confuse things. Now imagine that someone pushes the refrigerator and the stool with exactly the same force. The stool, being less massive than the refrigerator will obviously budge more easily and pick up speed more quickly.

What happens, however, if the stool and the refrigerator are acted on by the force of gravity? Say someone tips them both off the roof of a 10-story building? In this case, as Galileo himself would have predicted, the stool will not pick up speed faster than the refrigerator. Despite their wildly different masses, the stool and the refrigerator will accelerate towards the ground at exactly the same rate.

Now, perhaps you appreciate the central peculiarity of gravity. A big mass experiences a bigger force of gravity than a small mass, and that force is in direct proportion to its mass, so the big mass accelerates at exactly the same rate as the small mass. But how does gravity adjust itself to the mass it is acting on? It was Einstein's genius to realise that it does so in an incredibly simple and natural way—a way, furthermore, that has profound implications for our picture of gravity.

## THE EQUIVALENCE OF GRAVITY AND ACCELERATION

Say an astronaut is in a room accelerating upwards at 9.8 metres per second per second, which is the acceleration gravity imparts to falling bodies near Earth's surface. Think of the room as a cabin in a space-craft whose rocket engines have just started firing. Now, say the astronaut takes a hammer and a feather, holds them out from him at the same height above the floor of the cabin, then lets them go simultaneously. What happens? Well, the hammer and feather meet the floor of course. How this event is interpreted, though, depends entirely on the particular viewpoint.

Assuming the spacecraft is far away from the gravity of any big masses like planets, the hammer and the feather are weightless. So if we look into the spacecraft from outside with some kind of X-ray vision, we see the two objects hanging motionless. However, because the spacecraft is accelerating upward, we see the floor of the cabin racing up to meet the hammer and the feather. When it strikes them, furthermore, it strikes them both simultaneously.

Say the astronaut has amnesia and has entirely forgotten he is in a spacecraft. The portholes, in addition, are blacked out so there is nothing to tell him where he is. How does he interpret what he sees?

Well, the astronaut maintains that the hammer and the feather have fallen under gravity. After all, they have done the one thing a hammer and a feather experiencing gravity would do—they have fallen at the same rate and hit the ground at the same time (ignoring air resistance of course). The astronaut is further convinced that gravity is responsible for what he has seen by the fact that his feet appear to be glued to the floor just as they would be if he was in a room on Earth's surface. In fact, everything the astronaut experiences turns out to be indistinguishable from what he would experience if he was on Earth's surface.

Of course, it could be a coincidence. Einstein, however, was convinced he had stumbled onto a deep truth about nature. Gravity is indeed indistinguishable from acceleration, and the reason for that could not be simpler. Gravity is acceleration! This realisation, which Einstein later called "the happiest thought of my life," convinced him that the search for a theory of gravity and for a theory that described accelerated motion were one and the same thing.

Einstein elevated the indistinguishability of gravity and acceleration to a grand principle of physics, which he christened the principle
of equivalence. The principle of equivalence recognises that gravity is
not like other forces. In fact, it is not even a real force. We are all like
the amnesiac astronaut in the blacked-out spacecraft. We do not realise that our surroundings are accelerating and so have to find some
other way to explain away the fact that rivers flow downhill and apples
fall from trees. The only way is to invent a fictitious force—gravity.

## THE FORCE OF GRAVITY DOES NOT EXIST!

The idea that gravity is a fictitious force may sound a little far-fetched.
However, in other everyday situations, we are perfectly happy to invent forces to make sense of what happens to us. Say you are a passenger in a car that is racing round a sharp corner in the road. You
appear to be flung outward and, to explain why, you invent a force—
centrifugal force. In reality, however, no such force exists.

All massive bodies, once set in motion, have a tendency to keep
travelling at constant speed in a straight line.[1] Because of this property, known as inertia, unrestrained objects inside the car, including a
passenger like you, continue to travel in the same direction the car
was travelling before it rounded the bend. The path followed by the
car door however, is a curve. It should be no surprise, then, that you
find yourself jammed up against a door. But the car door has merely
come to meet you in the same way that the floor of the accelerating
spacecraft came up to meet the hammer and feather.[2] There is no
force.

---

[1]This is not at all obvious on Earth, where frictional forces act to slow a
moving body. However, it is apparent in the empty vacuum of space.

[2]It is worth pointing out that acceleration does not just mean a change
in speed. It can also mean a change in direction. So a car travelling around a
bend—even at constant speed—is accelerating.

Centrifugal force is known as an inertial force. We invent it to explain our motion because we choose to ignore the truth—that our surroundings are moving relative to us. But, really, our motion is just a result of our inertia, our natural tendency to keep moving in a straight line. It was Einstein's great insight to realise that gravity too is an inertial force. "Can gravitation and inertia be identical?" asked Einstein. "This question leads directly to my theory of gravity."

According to Einstein, we concoct the force of gravity to explain away the motion of apples falling from trees and planets circling the Sun because we ignore the truth—that our surroundings are accelerating relative to us. In reality, things move merely as a result of their inertia. The force of gravity does not exist!

But wait a minute. If the motion we attribute to the force of gravity is actually just the result of inertia, that must mean that bodies like Earth are really just flying through space at constant speed in straight lines. That's patently ridiculous! Earth is circling the Sun and not flying in a straight line, right? Not necessarily. It all depends on how you define a straight line.

## GRAVITY IS WARPED SPACE

A straight line is the shortest path between two points. This is certainly true on a flat piece of paper. But what about on a curved surface—for instance, the surface of Earth? Think of a plane flying the shortest route between London and New York. What path does it take? To someone looking down from space, it is obvious—a curved path. Think of a hiker trekking between two points in a hilly landscape. What path does the hiker take? To someone looking down on the hiker from a vantage point so high that the undulations of the landscape cannot be seen, the path of the hiker wiggles back and forth in the most tortuous manner.

Contrary to expectations, then, the shortest path between two points is not always a straight line. In fact, it is only a straight line on a very special kind of surface—a flat one. On a curved surface like

Earth's, the shortest route between two points is always a curve. In light of this point, mathematicians have generalised the concept of a straight line to include curved surfaces. They define a geodesic to be the shortest path between two points on any surface, not just a flat one.

What has all this got to do with gravity? The connection, it turns out, is light. It is a characteristic property of light that it always takes the shortest route between two points. For instance, it takes the shortest path from these words you are reading to your eyes.

Now think back to the amnesiac astronaut in his accelerating, blacked-out spacecraft. Bored of experimenting with a hammer and feather, he gets out a laser and places it on a shelf on the left-hand wall of his cabin, at a height of say 1.5 metres. He then crosses to the right-hand wall of the cabin and, with a marker pen, draws a red line, also at a height of 1.5 metres. Finally, the astronaut turns on the laser so that its beam stabs horizontally across the cabin. Where does it strike the right-hand wall?

It stands to reason that, since the astronaut has fired the beam horizontally, it will hit the wall exactly on the red line. So does it? The answer is *no!*

While the light is in flight across the cabin, the floor of the spacecraft is all the time being boosted by the rocket motors. Consequently, the floor is moving steadily upward to meet the beam. As the light gets closer and closer to the right-hand wall, the floor gets closer and closer to the light. Or from the point of view of the astronaut, the light gets closer and closer to the floor. Clearly, when the beam hits the right-hand wall, it hits it below the red line. The astronaut sees the light beam curving steadily downward as it crosses the cabin.

Now light, remember, always takes the shortest path between two points. The shortest path on something that is flat is a straight line, whereas the shortest path on something that is curved is a curve. What then are we to make of the fact that the light beam follows a curved trajectory across the spacecraft cabin? There is only one possible interpretation: The space inside the cabin is in some sense curved.

Now, you can argue that this is just an illusion caused by the accelerating spacecraft. The crucial point, however, is that the astronaut has no way of knowing that he is in an accelerating spacecraft. He could just as well be experiencing gravity in a room on Earth's surface. Acceleration and gravity are indistinguishable. This is the principle of equivalence. What the experiment with the laser beam is actually demonstrating—and this shows the tremendous power of the principle of equivalence—is that light in the presence of gravity follows a curved trajectory. Or to put it another way, gravity bends the path of light.

Gravity bends light because space, in the presence of gravity, is somehow curved. In fact, this is all gravity turns out to be—curved space.

What exactly do we mean by curved space? It is easy to visualise a curved surface like the surface of Earth. But that is because it has only two directions, or dimensions—north-south and east-west. Space is a bit more complicated than that. In addition to three space dimensions—north-south, east-west, and up-down—there is one time dimension—past-future. As Einstein showed, however, space and time are really just aspects of the same thing, so it is more accurate to think of there being four "space-time" dimensions.

Four-dimensional space-time is impossible for us to visualise since we live in a world of three-dimensional objects. This means that the curvature, or warpage, of four-dimensional space-time is doubly impossible to visualise. But that's what gravity is: the warpage of four-dimensional space-time.

Fortunately, we can get some idea of what this means. Imagine a race of ants that spends its entire existence on the two-dimensional surface of a taut trampoline. The ants can only see what happens on the surface and have no concept whatsoever of the space above and below the trampoline—the third dimension. Now imagine that you or I—mischievous beings from the third dimension—put a cannonball on the trampoline. The ants discover that when they wander near the cannonball their paths are mysteriously bent towards it. Quite reasonably, they explain their motion by saying that the cannonball is exerting a force of attraction on them. Perhaps they even call the force gravity.

However, from the God-like vantage point of the third dimension, it is clear the ants are mistaken. There is no force attracting them to the cannonball. Instead, the cannonball has made a valleylike depression in the trampoline, and this is the reason the paths of the ants are bent towards it.

Einstein's genius was to realise that we are in a remarkably similar position to the ants on the trampoline. The path of Earth as it travels through space is constantly bent towards the Sun, so much so that the planet traces out a near-circular orbit. Quite reasonably, we explain away this motion by saying that the Sun exerts a force of attraction on Earth—the force of gravity. However, we are mistaken. If we could see things from the God-like perspective of the fourth dimension—something that is as impossible for us to do as it is for the ants to see things from the third dimension—we would see there is no such force. Instead, the Sun has created a valleylike depression in the four-dimensional space-time in its vicinity, and the reason Earth follows a near-circular path around it is because this is the shortest possible path through the warped space.

There is no force of gravity. Earth is merely following the straightest possible line through space-time. It is because space-time near the Sun is warped that that line happens to be a near-circular orbit. According to physicists Raymond Chiao and Achilles Speliotopoulos: "In general relativity, no 'gravitational force' exists. What we normally associate with the force of gravity on a particle is not a force at all: The particle is simply travelling along the 'straightest' possible path in curved space-time."

A body travelling along the "straightest" possible path through space-time is in free fall. And, since it is in free fall, it experiences no gravity. Earth is in free fall around the Sun. Consequently, we do not feel the Sun's gravity on Earth. The astronauts on the International

Space Station are in free fall around Earth. Consequently, they do not feel Earth's gravity.[3]

Gravity arises only when a body is prevented from following its natural motion. Our natural motion is free fall towards the centre of Earth. The ground thwarts us, however, so we feel its force on our bodies, which we interpret as gravity. Just as centrifugal force is what we feel when a cornering car prevents us from following the natural motion in a straight line, the force of gravity is what we feel when our surroundings prevent us from following our natural motion along a geodesic.

Probably, it seems unnecessarily complicated to view massive bodies as moving under their own inertia through warped space-time rather than simply moving under the influence of a universal force of gravitational attraction. However, the two pictures are not equivalent. Einstein's is superior. For a start, the thing that is warped is not merely space but the space-time of special relativity. The picture, therefore, automatically incorporates the peculiar interplay between space and time necessary to keep the speed of light a constant. Einstein's picture also predicts new things.

Think of those ants on the trampoline. There are more things you can do with the material of the trampoline than merely depress it with a heavy mass like a cannonball. For instance, you could shake one corner up and down. This would cause ripples in the fabric to spread outwards across the trampoline like ripples on the surface of a

---

[3]Most people assume that astronauts orbiting Earth are weightless because there is no gravity in space. However, at the 500-kilometre-or-so height of the International Space Station, gravity is only about 15 per cent weaker than on Earth's surface. The real reason astronauts are weightless is that they and their spacecraft are in free fall just as surely as someone in an elevator when the cable breaks. The difference is that they never hit the ground. Why? Because Earth is round and, as fast as they fall toward the surface, the surface curves away from them. They, therefore, fall forever in a circle.

pond. In the same way, the vibration of a large mass like a black hole out in space can generate ripples in the "fabric" of space-time. Such gravitational waves have yet to be detected directly, but their existence is a unique prediction of Einstein's theory.

The fact that waves can ripple through space-time suggests that space is not the empty, passive medium imagined by Newton. Instead, it is an active medium with real properties. Matter does not simply pull on other matter across empty space, as Newton imagined. Matter distorts space-time, and it is this distorted space-time that in turn affects other matter. As John Wheeler put it: "Mass tells space-time how to warp and warped space-time tells mass how to move."

The distortion of space-time caused by a massive body takes time to propagate to another mass, just as the distortion of the trampoline by another cannonball takes time to reach the corners of the trampoline. Because of this, gravity—warped space-time—acts only after a delay, in perfect accord with the cosmic speed limit set by the speed of light.

The fact that space-time has some of the qualities of a real medium like air or water has implications for large bodies like planets and stars. When they rotate on their axes, they actually drag space-time around with them. NASA has measured the effect, known as frame dragging, with an orbiting space experiment called Gravity Probe B. Frame dragging is tiny in the case of Earth but overwhelming in the case of a rapidly spinning black hole. Such a body sits at the eye of a great tornado of spinning space-time. Anyone falling into the black hole would be whirled around with the tornado, which no power in the Universe could defy.

## THE RECIPE OF GENERAL RELATIVITY

Einstein's novel take on gravity is now clear. Masses—for instance, stars like the Sun—warp the space-time around them. Other masses—for instance, planets like Earth—then fly freely under their own inertia through the warped space-time. The paths they follow are curved because these are the shortest possible paths in warped space. This is it. This is the general theory of relativity.

The devil, however, is in the details. We know how a massive body like a planet moves in warped space. It takes the shortest possible path. But how precisely does a mass like the Sun warp the space-time around it? It took Einstein more than a decade to find out, and the details would fill a textbook as big as a phone directory. However, Einstein's starting point for the general theory of relativity is not difficult to appreciate. It is none other than the principle of equivalence.

Recall again the hammer and the feather in the blacked-out spacecraft. To the astronaut, they appeared to fall to the floor under gravity. To someone watching the experiment from outside the spacecraft, however, it was obvious that the hammer and the feather were hanging in midair and that the floor of the cabin was accelerating upwards to meet them. They were completely weightless.

This observation is of key importance. A body falling freely in gravity feels no gravity. Imagine you are in an elevator and someone cuts the cable. As it falls, you are weightless; you feel no gravity.

"The breakthrough came suddenly one day," Einstein wrote in 1907. "I was sitting on a chair in my patent office in Bern. Suddenly the thought struck me: If a man falls freely, he does not feel his own weight. I was taken aback. This simple thought experiment made a deep impression on me. This led me to the theory of gravity."

What is the significance of a freely falling body feeling no gravity? Well, if it experiences no gravity—or acceleration, since the two are the same—then its behaviour is described entirely by Einstein's special theory of relativity. Here then is the crucial point of contact—the all-important bridge—between the special theory of relativity and the theory of gravity sought by Einstein.

The observation that a freely falling body does not feel its weight and is therefore described by special relativity suggests a crude way to extend special relativity to a body experiencing gravity. Think of a friend standing on Earth and very obviously experiencing gravity pressing his or her feet to the ground. You can observe your friend from any point of view you like—from hanging upside down from a nearby tree or from an aeroplane flying past. But one point of view provides a big payoff. If you imagine things from a point of view that is in free fall, then you will be weightless, subject to no acceleration. Since you feel no acceleration, you are justified in using the special theory of relativity to describe your friend.

But special relativity relates what the world looks like to people moving at constant speed relative to each other and your friend is accelerating upwards relative to you. That's true. But if you do not mind a lot of laborious calculation, you can imagine your friend travelling at constant speed, a second, say then at a slightly higher constant speed for the next second, and so on. It's not perfect, but you can approximate your friend's acceleration as a series of rapid steps up in speed. For each speed you simply use special relativity to tell you what is happening to the space and time of your friend.

According to special relativity, time slows down for a moving observer. It therefore follows that time slows down for your friend since your friend is moving relative to you. But wait. Your friend is moving relative to you because he or she is experiencing gravity. It follows that gravity must slow down time! This should not be too much of a surprise. After all, if gravity is simply the warpage of space-time, it stands to reason that if we are experiencing gravity, our space and our time must be distorted in some way.

The other thing that follows from thinking about your friend standing on Earth's surface is that if gravity were stronger—say your friend was standing on a more massive planet—his or her speed relative to you in free fall would get faster quicker. According to special relativity, the faster someone moves, the more their time slows down. Consequently, the stronger the gravity someone is experiencing, the more their time slows down. What this means is that if you work on the ground floor of an office building, you age more slowly than your colleagues who work on the top floor. Why? Because, being closer to

Earth, you experience a more powerful pull, and time slows down in stronger gravity.

Earth's gravity, however, is very weak. After all, you can hold your arm out in front of you and not even the gravity of the entire Earth can force you to drop it. The weakness of Earth's gravity means that the difference in the flow rate of time between the ground and top floors of even the tallest building is nearly impossible to measure. The opening scene, with the twin sisters aging at vastly different rates in their skyscraper workplace, is therefore a gross exaggeration. Nevertheless, there are places in the Universe with far stronger gravity.

One place is the surface of a white dwarf star, where the gravity is much stronger even than the Sun's. Einstein's theory of gravity predicts that time for these stars should pass slightly slower than for us. Testing such a prediction might seem impossible. However, nature has very conveniently provided us with "clocks" on the surfaces of white dwarfs. The clocks are actually atoms.

Atoms give out light. Light is actually a wave that undulates up and down like a wave on water, and atoms of particular elements such as sodium or hydrogen give out light that is unique to the element, undulating a characteristic number of times a second. These undulations can be thought of as the ticks of a clock. (In fact, the second is defined in terms of the undulations of light given out by a particular type of atom.)

How does this property of atoms help us see the effect of gravity on time? Well, with our telescopes we can pick up the light from atoms on white dwarfs. We can then compare the number of undulations per second of the light from, say, hydrogen on a white dwarf, with the number of undulations per second of hydrogen on Earth. What we find is that there are fewer undulations per second in the light from a white dwarf. Light is more sluggish. Time runs slower![4]

---

[4]For technical reasons, this effect is known as the gravitational red shift.

We are seeing a direct confirmation of Einstein's general theory of relativity.

And there are stars known as neutron stars with even stronger gravity than that of white dwarfs. As a result of the strong gravity, time on the surface of a neutron star progresses one and a half times more slowly than on Earth.

## THE CONSEQUENCES OF GENERAL RELATIVITY

Time dilation is only one of the novel predictions of Einstein's general theory of relativity. Another, already touched on, is the existence of gravitational waves. We know they exist because astronomers have observed pairs of stars, which include at least one neutron star, losing energy as they spiral in towards each other. This puzzling loss of energy can be explained only if it is being carried away by gravitational waves.

The race is now on to detect gravitational waves directly. As they pass by, they should alternately stretch and squeeze space. Experiments designed to detect them therefore use giant "rulers," many kilometres long. The rulers are made of light, but the idea is simple—to detect the change in length of the rulers as a gravitational wave ripples past.

Another prediction of Einstein's theory, so far passed over without comment, is the bending of light by gravity. The reason for this bending, of course, is that light must negotiate the warped terrain of four-dimensional space-time. Although Newton's law of gravity predicts no such effect, it does when combined with the special relativistic idea that all forms of energy—including light—have an effective mass. As light passes a massive body like the Sun, it therefore feels the tug of gravity and is bent slightly from its course.

Of course, special relativity is incompatible with Newton's law of gravity, so this light-bending prediction has to be taken with a pinch of salt. In fact, the correct theory—general relativity—predicts that the path of light will be bent by twice as much.

This extra factor of two serves to highlight something subtle about the principle of equivalence. Recall the experiment in which the astronaut fired the laser horizontally across his spacecraft and noticed that the beam was bent downwards. Because there was no way he could know he was not experiencing gravity in a room on Earth's surface, it was possible to deduce that gravity bends the path of light. Well, there is a little lie in here. You see, it turns out that it *is* possible for the astronaut to tell whether he is in a rocket or on Earth's surface.

In the accelerating rocket, the force that pins the astronaut's feet to the floor pulls him vertically downwards—wherever he stands in the cabin. On Earth's surface, however, it matters where you stand because gravity always pulls things towards the centre of Earth. Consequently, gravity pulls in one direction in England but in the opposite direction in New Zealand—to the English, the New Zealanders are upside down, and vice versa. Now, the direction of the pull of gravity does not change too much from one side of a room to another. Nevertheless, with sensitive-enough measuring instruments, our astronaut could always detect the change and tell whether he was in a rocket accelerating out in space or on Earth's surface.

Surely, this invalidates the principle of equivalence and brings the whole edifice of general relativity tumbling down? Well, you might think so. However, to construct a theory of gravity it is sufficient only that the principle of equivalence apply in tiny volumes of space, and in extremely tiny, localised volumes of space you can never detect changes in the direction of gravity.

What has this got to do with Einstein's theory predicting twice the light deflection of Newton's? Well, we have established that the laser beam will be bent downwards as it traverses a room on Earth's surface, and this amount turns out to be roughly what Newtonian gravity predicts. Now imagine that the room is in free fall—say it has been dropped from an aeroplane—and the astronaut carries out the same experiment. In free fall, remember, there is no gravity. So the light beam should travel horizontally across the room and not be bent at all. But not all parts of the room are in a perfect state of free fall. Because Earth's gravity pulls in one direction from one corner of the room and from a different direction from the other corner, gravity is not perfectly cancelled out as the room falls through the air. Because of this, what the astronaut actually sees is the light beam bent downwards by roughly the same amount as in the room on Earth's surface. The two effects add together to give twice the light bending predicted by Newton's theory of gravity plus special relativity.

So if the light from a distant star passes close to the Sun on its way to Earth, its trajectory should be bent about twice as sharply as Newton would have predicted. Such an effect would cause the position of a star to shift slightly relative to other stars. Though impossible to see in the glare of daylight, it is observable during a total eclipse when the Moon blots out the bright solar disc. Such an eclipse was due to occur on May 29, 1919, and the English astronomer Arthur Eddington travelled to the island of Principe off the coast of West Africa to see it. His photographs confirmed that starlight was indeed deflected by the Sun's gravity by exactly the amount predicted by the general theory of relativity.

Eddington's observations made Einstein's reputation as "the man who proved Newton wrong." But it was not the end of general relativity's successful predictions. Newton had demonstrated theoretically that the planets orbited the Sun not in circles but in ellipses— squashed circles. He proved that this was a direct consequence of the fact that the force of gravity drops off in strength with a so-called inverse-square law. In other words, when you are twice as far away from the Sun, the force of gravity is four times as weak; three times as far away, it is nine times as weak; and so on.

Relativity changes everything. For a start, all forms of energy, not just mass-energy, generate gravity. Now gravity itself is a form of energy. Think of a warped trampoline and how much elastic energy that contains. Since gravity is a form of energy, the gravity of the Sun itself creates gravity! It's a tiny effect and most of the Sun's gravity still comes from its mass. Nevertheless, close in to the Sun, where gravity is strong, there is a small extra contribution from gravity itself. Consequently, any body orbiting there feels a gravitational tug greater than expected from the inverse square law.

Now—and this is the point—planets follow elliptical orbits only if they are being tugged by a force obeying an inverse-square law of force. This was Newton's discovery. Relativity predicts that the force does not obey an inverse-square law. In fact, there are other effects that also cause a departure from Newtonian gravity, like the fact that gravity takes time to travel across space. The gravity that a moving planet feels at any moment therefore depends on its position at an earlier time and, because of this, is not directed towards the dead centre of the Sun. The upshot is that planets do not follow elliptical paths that repeat but rather elliptical paths which gradually change their orientation in space, tracing out a rosette-like pattern. This is not noticeable far from the Sun. The biggest effect is close in, where gravity is strongest.

Sure enough, there is something odd about the orbit of the innermost planet, Mercury. For some time before Einstein published his theory of gravity in 1915, astronomers had been puzzled by the fact that Mercury's orbit gradually traces out a rosette pattern in space. Most of this effect is due to the gravitational pull of Venus and Jupiter. The odd thing, however, is that Mercury's orbit would still be tracing out a rosette pattern *even if Venus and Jupiter were not there*. It is a tiny effect. Although Mercury orbits the Sun once every 88 days, a rosette is traced out only once every *3 million years*. Remarkably, this is exactly what Einstein's theory predicted. Using general relativity, he could explain every last detail of Mercury's orbit. With yet another successful prediction under its belt, there could be no doubt that Einstein had discovered the correct theory of gravity.[5]

---

[5]Or at least a workable theory for the time being, since even general relativity is not thought to be the last word on gravity.

## THE PECULIARITIES OF GENERAL RELATIVITY

General relativity is a fantastically elegant theory. Nevertheless, it is tremendously difficult to apply to real situations—for instance, to find the warpage of space-time caused by a given distribution of mass. The reason is that the theory is rather circular. Matter tells space-time how to warp. Then warped space-time tells matter how to move. The matter, which has just moved, tells space-time how to change its warpage. And so on, ad infinitum. There's a kind of chicken-and-egg paradox at the heart of the theory. Physicists call it nonlinearity, and nonlinearity is a tough nut for theorists to crack.

One manifestation of nonlinearity already mentioned is the fact that gravity is a source of gravity. Well, if gravity can make more gravity, that extra gravity can make a little more gravity, and so on. Fortunately, gravity is so weak that this is not normally a runaway process and the gravity generated by a massive body is usually well behaved—usually, but not always.

Some very massive stars end their lives in a spectacular way. Usually, a star is prevented from being crushed by its own gravity by the pressure of the hot gas in its interior pushing outwards. But this outward pressure only exists while the star is generating heat. When it runs out of all possible fuels, it shrinks. Usually, some other form of pressure intervenes to make a white dwarf or a neutron star, superdense stellar embers. However, if the star is very massive and its gravity is very strong, nothing can stop the star from shrinking down to a point. As far as physicists know, such stars literally vanish from existence. However, they leave something behind: their gravity.

What we are talking about here are black holes, perhaps the most bizarre of all the predictions of general relativity. A black hole is a region of space-time where gravity is so strong that not even light can escape it—hence its blackness. And "region of space-time" is the operative phrase, for the mass of the star has gone.

How can you have gravity without mass? Well, gravity arises not just from mass but from all forms of energy. In the case of the black hole, its own gravity creates more gravity and that extra gravity creates more gravity . . . so the hole regenerates itself like a man holding himself in midair by his boot straps. From the space-time point of view, a black hole is literally a hole. Whereas a star like the Sun creates a mere dimple in the surrounding space-time, a black hole produces a bottomless well into which matter falls but can never escape again.

As Nobel Prize-winning physicist Subrahmanyan Chandrasekhar observed: "The black holes of nature are the most perfect macroscopic objects there are in the universe: The only elements in their construction are our concepts of space and time."[6]

Because of their ultrastrong gravity, black holes reveal the most dramatic effects of general relativity. Surrounding them is a surface known as an event horizon. This marks the point of no return for objects straying too close to the black hole. If you moved in close to the event horizon, you could see the back of your head since light from behind you would be bent all the way around the hole before reaching your eyes. If you could somehow hover just outside the event horizon, time would flow so slowly for you that you could in theory watch the entire future of the Universe flash past you like a movie in fast-forward!

The fact that time runs far more slowly in the strong gravity of a black hole than elsewhere in the Universe has an intriguing consequence. Imagine you are far away from a black hole and you have a friend lingering close to it. Because of the marked difference in the flow of time for both of you, while you go from Monday to Friday,

---

[6]The term "black hole" was coined by John Wheeler in 1965. Before 1965 there were very few scientific papers on such objects. Afterward, the field exploded. The term has even entered everyday language. People often talk about things disappearing down a bureaucratic black hole. The term is a perfect illustration of the importance of getting the right words to describe a phenomenon in science. If they paint a vivid picture in people's minds, researchers are attracted to the subject.

your friend progresses only from Monday to Tuesday. This means that, if you could find some way to spirit yourself over to your friend's location, you could go from Friday back to Tuesday. You could travel back in time!

It turns out that there is in fact a way to spirit yourself from one location to another. Einstein's theory of relativity permits the existence of "wormholes," tunnel-like shortcuts through space-time. By entering one mouth of such a wormhole and exiting a mouth near your friend, it would indeed be possible to go back in time from Friday to Tuesday.

The trouble with wormholes is that they snap shut in an instant unless held open by matter with repulsive gravity. Nobody knows whether such "exotic matter" exists in the Universe. Nevertheless, the extraordinary fact remains that Einstein's theory of gravity does not rule out the possibility of time travel.

There are a few differences, however, between the kind of "time machine" permitted by general relativity and the type described by science fiction writers like H. G. Wells. For one thing, you have to travel a distance through space to travel a distance through time. You cannot simply sit still in a time machine, pull a lever, and find yourself in 1066. And a second important difference is that you cannot go back to a time before your time machine was built. So if you want to go on a dinosaur safari, building a time machine today will not help. You will have to find one built and abandoned by extraterrestrials (or some very smart dinosaurs) 65 million years ago!

To theorists the possibility of time machines is very unsettling. If time travel is possible, all sorts of impossible situations, or "paradoxes," raise their ugly heads. The most famous is the grandfather paradox in which a man goes back in time and shoots his grandfather before he conceives the man's mother. The problem is, if he shoots his grandfather, how can he ever be born to go back in time and do the dirty deed?!

Embarrassing questions like this have prompted the English physicist Stephen Hawking to propose the Chronology protection conjecture. Basically, it's just a fancy name for an outright ban on time travel. According to Hawking, some as-yet-unknown law of physics must intervene to prevent time travel. He has no cast-iron evidence of such a law but simply asks: "Where are the tourists from the future?"

Einstein himself did not believe that time travel was possible, despite the fact that his theory of gravity predicted it. He was wrong, however, about two other predictions of his theory. He did not believe that black holes were possible, and today we have compelling evidence that they exist. And he did not believe what his theory was trying to tell him about the origin of the Universe—that it began in a Big Bang.

# 10

# THE ULTIMATE RABBIT
# OUT OF A HAT

HOW WE LEARNED THAT THE UNIVERSE HAS NOT EXISTED FOREVER BUT WAS BORN
IN A TITANIC EXPLOSION **13.7** BILLION YEARS AGO

*A white rabbit is pulled out of a top hat. Because it is an extremely
large rabbit, the trick takes billions of years.*

Jostein Gaarder

*They are high-tech glasses. Merely by twiddling a knob on the frame,
you can "tune" them to see all kinds of light normally invisible to the
human eye. You take them outside on a cold, starry night and start twiddling.*

*The first thing you see is the sky in ultraviolet, light pumped out by
stars much hotter than the Sun. Some familiar stars have vanished, and
some new ones have swum into view, shrouded in misty nebulosity. The
most striking feature of the sky, however, is the same as it was for the
naked-eye sky. It's mostly black.*

*You twiddle on.*

*Now you're seeing X-rays, high-energy light radiated by gas heated
to hundreds of thousands of degrees as it swirls down onto exotic objects
like black holes. Once again, the most striking feature of the sky is that it
is mostly black.*

*You twiddle back the other way, zipping back through ultraviolet
light and visible light to infrared light, given out by objects much colder
than the Sun. Now the sky is peppered by stellar embers—stars so*

*recently born they are still swathed in shimmering placental gas and bloated red giants in their death throes. But despite the fact that the sky is lit by a new population of stars, its most striking thing remains the same. It is mostly black.*

*You twiddle on. Now you are seeing microwaves—the kind of light used for radar, mobile phones, and microwave ovens. But something odd is happening. The sky is getting brighter. Not just bits of it—all of it!*

*You take off the glasses, rub your eyes, and put them back on. But nothing has changed. Now the whole sky, from horizon to horizon, is glowing a uniform, pearly white. You twiddle further, but the sky just gets brighter and brighter. The whole of space seems to be glowing. It's like being inside a giant lightbulb.*

Are the glasses malfunctioning? No, they are working perfectly. What you are seeing is the cosmic background radiation, the relic of the fireball in which the Universe was born 13.7 billion years ago. Incredibly, it still permeates every pore of space, greatly cooled by the expansion of the Universe so that it now appears as low-energy microwaves rather than visible light. Believe it or not, the cosmic background radiation accounts for an astonishing 99 per cent of the light in today's Universe. It is incontrovertible proof that the Universe began in a titanic explosion—the Big Bang.

The cosmic background radiation was discovered in 1965. But the realisation that there had been a Big Bang actually came earlier. In fact, the first step was taken by Einstein.

## THE ULTIMATE SCIENCE

Einstein's theory of gravity—the general theory of relativity—describes how every chunk of matter pulls on every other chunk of matter. The biggest collection of matter we know of is the Universe. Never one to shy away from the really big problems in science, Einstein in 1916 applied his theory of gravity to the whole of creation. In doing so he created cosmology—the ultimate science—which deals with the origin, evolution, and ultimate fate of the Universe.

Although the ideas behind Einstein's theory of gravity are deceptively simple, the mathematical apparatus is not. Working out exactly how a particular distribution of matter warps space-time is very hard indeed. It was not until 1962, for instance—almost half a century after Einstein published his general theory of relativity—that New Zealand physicist Roy Kerr calculated the distortion of space-time caused by a realistic, spinning, black hole.

Figuring out how the whole Universe warps space-time would have been impossible without making some simplifying assumptions about how its matter is spread throughout space. Einstein assumed that it makes no difference where in the Universe an observer happens to be. In other words, he assumed that the Universe has the same gross properties wherever you are located and, from wherever you are located, it looks roughly the same in every direction.

Astronomical observations since 1916 have actually shown these assumptions to be well founded. The Universe's building blocks—which Einstein and everyone else were unaware of at the time—are galaxies, great islands of stars like our own Milky Way. And modern telescopes do indeed show them to be scattered pretty evenly around the Universe, so the view from one galaxy is much the same as the view from any other.

Einstein's conclusion, after applying his theory to the Universe as a whole, was that its overall space-time must be warped. Warped space-time, however, causes matter to move. This is the central mantra of general relativity. Consequently, the Universe could not possibly be still. This dismayed Einstein. Like Newton before him, he fervently believed the Universe to be static, its constituent bodies—now known to be galaxies—suspended essentially motionless in the void.

A static universe was appealing because it remained the same for all time. There was no need to address sticky questions about where the Universe came from or where it was going. It had no beginning. It had no end. The reason the Universe was the way it was was because that was the way it had always been.

According to Newton, for the Universe to be static, one condition had to be satisfied: matter had to extend infinitely in all directions. In such a neverending cosmos, each body has just as many bodies on one side, pulling it one way with their gravity, as on the opposite side, pulling it the other way. Like a rope being pulled by two equally strong tug-of-war teams, it therefore remains motionless.

However, according to Einstein's theory of gravity, the Universe was finite, not infinite. Its space-time curved back on itself—the four-dimensional equivalent of the two-dimensional surface of a basketball. In such a Universe the gravitational tug-of-war is at no point perfectly balanced. Because every body tries to pull every other body toward it, the Universe shrinks uncontrollably.

To salvage the idea of a static Universe, Einstein had to resort to mutilating his elegant theory. He added a mysterious force of cosmic repulsion, which pushed apart the objects in the Universe. He hypothesised that it had a significant effect only on bodies that were enormously far apart, explaining why it had not been noticed before in Earth's neighbourhood. By precisely counteracting the force of gravity that was perpetually trying to drag bodies together, the cosmic repulsion kept the Universe forever static.

## THE EXPANDING UNIVERSE

Einstein's instincts turned out to be wrong. In 1929, Edwin Hubble—the American astronomer responsible for discovering that the Universe's building blocks were galaxies—announced a dramatic new discovery. The galaxies were flying apart from each other like pieces of cosmic shrapnel. Far from being static, the Universe was growing in size. As soon as Einstein learned of Hubble's discovery of the expanding universe, he renounced his cosmic repulsion, calling it the

biggest blunder he ever made in his life.[1] Einstein's mysterious repulsive force could never have kept the galaxies hanging motionless
in space. As Arthur Eddington pointed out in 1930, a static cosmos is
inherently unstable, like a knife balanced on its point. The merest
nudge would be enough to set it expanding or contracting.

Others did not make the same mistake as Einstein. In 1922 the
Russian physicist Aleksandr Friedmann applied Einstein's theory of
gravity to the Universe and correctly concluded that it must either be
contracting or expanding. Five years later the same conclusion was
reached independently by the Belgian Catholic priest Georges-Henri
Lemaître.

As John Wheeler has said: "Einstein's description of gravity as
curvature of space-time led directly to that greatest of all predictions:
The Universe itself is in motion." It is ironic that Einstein himself
missed the message in his own theory.

## THE BIG BANG UNIVERSE

Since the Universe is expanding, one conclusion is inescapable: it
must have been smaller in the past. By imagining the expansion running backwards, like a movie in reverse, astronomers deduce that 13.7
billion years ago all of Creation was squeezed into the tiniest of tiny
volumes. The lesson of the receding galaxies is that the Universe,
though old, has not existed forever. There was a beginning to time. A
mere 13.7 billion years ago, all matter, energy, space, and time
fountained into existence in a titanic explosion—the Big Bang.

The cosmic expansion turns out to obey a remarkably simple
law: Every galaxy is rushing away from the Milky Way with a speed
that is in direct proportion to its distance. So a galaxy that is twice as

---

[1] See *My World Line* by George Gamow (New York, 1970), in which the
author writes of Einstein: "He remarked [to me] that the introduction of the
cosmological term was the biggest blunder he ever made in his life."

far away as another is receding twice as fast, one 10 times as far away 10 times as fast, and so on. This relation, known as Hubble's law, turns out to be unavoidable in any universe that grows in size while continuing to look the same from every galaxy.

Imagine a cake with raisins in it. If you could shrink in size and sit on any raisin, the view will always be the same. Furthermore, if the cake is put in an oven and expands, or rises, not only will you see all the other raisins recede from you but you will see them recede with speeds in direct proportion to their distance from you. It matters not at all what raisin you sit on. The view will always be the same. (The tacit assumption here is that it is a big cake, so that you are always far from the edge.) Galaxies in an expanding universe are like raisins in a rising cake.

It follows that, just because we see all the galaxies flying away from us, we should not assume that we are at the centre of the Universe and that the Big Bang happened in our cosmic backyard. Were we to be in any galaxy other than the Milky Way, we would see the same thing—all the other galaxies fleeing from us. The Big Bang did not happen here, or over there, or at any one point in the Universe. It happened in all places simultaneously. "In the universe, no centre or circumference exists, but the centre is everywhere," said the 16th-century philosopher Giordano Bruno.

The Big Bang is a bit of a misnomer. It was totally unlike any explosion with which we are familiar. When a stick of dynamite detonates, for instance, it explodes outwards from a localised point and the debris expands into preexisting space. The Big Bang did not happen at a single point and there was no preexisting void! Everything—space, time, energy, and matter—came into being in the Big Bang and began expanding everywhere at once.

## THE HOT BIG BANG

Whenever you squeeze something into a smaller volume—for instance, air into a bicycle pump—it gets hot. The Big Bang was therefore a hot Big Bang. The first person to realise this was the Ukrainian-American physicist George Gamow. In the first few moments after the Big Bang, he reasoned, the Universe was reminiscent of the blisteringly hot fireball of a nuclear explosion.[2]

But whereas the heat and light of a nuclear fireball dissipate into the atmosphere so that, hours or days after the explosion, they are all gone, this was not true of the heat and light of the Big Bang fireball. Since the Universe, by definition, is all there is, there was simply nowhere for it to go. The "afterglow" of the Big Bang was instead bottled up in the Universe forever. This means it should still be around today, not as visible light—since it would have been greatly cooled by the expansion of the Universe since the Big Bang—but as microwaves, an invisible form of light characteristic of very cold bodies.[3]

Gamow did not believe it would be possible to distinguish this microwave afterglow from other sources of light in today's Universe. However, he was mistaken. As his research students Ralph Alpher and Robert Herman realised, the relic heat of the Big Bang would have two unique features that would make it stand out. First, because it came from the Big Bang, and the Big Bang happened everywhere simultaneously, the light should be coming equally from every direction in the sky. And, second, its spectrum—the way the brightness of the light changed with the light's energy—would be that of a "black body." It's not necessary to know what a black body is, only that a black body spectrum is a unique "fingerprint."

Although Alpher and Herman predicted the existence of the afterglow of the Big Bang—the cosmic microwave background radiation—in 1948, it was not discovered until 1965 and then totally by

---

[2]The Big Bang was named by the English astronomer Fred Hoyle during a BBC radio programme in 1949. The great irony is that Hoyle, to the day he died, never believed in the Big Bang.

[3]And of magnetrons, which power microwave ovens and radar transmitters.

accident. Arno Penzias and Robert Wilson, two young astronomers at Bell Labs at Holmdel in New Jersey, were using a horn-shaped microwave antenna formerly used for communicating with *Telstar*, the first modern communications satellite, when they picked up a mysterious hiss of microwave "static" coming equally from every direction in the sky. Over the following months as they puzzled over the signal, they variously thought that it might be radio static from nearby New York City, atmospheric nuclear tests, or even pigeon droppings coating the interior of their microwave horn. In fact, they had made the most important cosmological discovery since Hubble found that the Universe was expanding. The afterglow of creation was powerful evidence that our Universe had indeed begun in a hot, dense state—a Big Bang—and had been growing in size and cooling ever since.

Penzias and Wilson did not accept the Big Bang origin of their mysterious static for at least two years. Nevertheless, for the discovery of the afterglow of creation, they carried off the 1978 Nobel Prize for Physics.

The cosmic background radiation is the oldest "fossil" in creation. It comes to us directly from the Big Bang, carrying with it precious information about the state of the Universe in its infancy, almost 13.7 billion years ago. The cosmic background is also the coldest thing in nature—only 2.7 degrees above absolute zero, the lowest possible temperature (−270 degrees Celsius).

The cosmic background radiation is actually one of the most striking features of our Universe. When we look up at the night sky, its most obvious feature is that it is mostly black. However, if our eyes were sensitive to microwave light rather than visible light, we would see something very different. Far from being black, the entire sky, from horizon to horizon, would be white, like the inside of a lightbulb. Even billions of years after the event, all of space is still glowing softly with relic heat of the Big Bang fireball.

In fact, every sugarcube-sized region of empty space contains 300 photons of the cosmic background radiation. Ninety-nine per cent of all the photons in the Universe are tied up in it, with a mere 1 per cent in starlight. The cosmic background radiation is truly ubiquitous. If you tune your TV between stations, 1 per cent of the "snow" on the screen is the relic static of the Big Bang.

## DARKNESS AT NIGHT

The fact that the Universe began in a Big Bang explains another great mystery—why the night sky is dark. The German astronomer Johannes Kepler, in 1610, was the first to realise this was a puzzle.

Think of a forest of regularly spaced pine trees going on forever. If you ran into the forest in a straight line, sooner or later you would bump into a tree. Similarly, if the Universe is filled with regularly spaced stars and goes on forever, your gaze will alight on a star no matter which direction you look out from Earth. Some of those stars will be distant and faint. However, there will be more distant stars than nearby ones. In fact—and this is the crucial point—the number of stars will increase in such a way that it exactly compensates for their faintness. In other words, the stars at a certain distance from Earth will contribute just as much light in total as the ones twice as far away, three times away, four times away, and so on. When all the light arriving at Earth is added up, the result will therefore be an infinite amount of light!

This is clearly nonsensical. Stars are not pointlike; they are tiny discs. So nearby stars blot out some of the light of more distant stars just as nearby pine trees block out more distant pine trees. But even taking this effect into account, the conclusion seems inescapable that the entire sky should be "papered" with stars, with no gaps in between. Far from being dark at night, the night sky should be as bright as the surface of a typical star. A typical star is a red dwarf, a star glowing like a dying ember. Consequently, the sky at midnight should be glowing blood red. The puzzle of why it isn't was popularised in the early 19th century by the German astronomer Heinrich Olbers and is known as Olbers' paradox in his honour.

The way out of Olbers' paradox is the realisation that the Universe has not in fact existed forever but was born in a Big Bang. Since the moment of creation, there has been only 13.7 billion years for the light of distant stars to reach us. So the only stars and galaxies we see are those that are near enough that their light has taken less than 13.7 billion years to get to us. Most of the stars and galaxies in the Universe are so far away that their light will take more than 13.7 billion years to reach us. The light of these objects is still on its way to Earth.

Therefore, the main reason the sky at night is dark is that the light from most of the objects in the Universe has yet to reach us. Ever since the dawn of human history, the fact that the Universe had a beginning has been staring us in the face in the darkness of the night sky. We have simply been too stupid to realise it.

Of course, if we could wait another billion years, we would see stars and galaxies so far away that their light has taken 14.7 billion years to get here. The question therefore arises of whether, if we lived many trillions of years in the future when the light from many more stars and galaxies had time to reach us, the sky at night would be red. The answer turns out to be *no*. The reasoning of Kepler and Olbers is based on an incorrect assumption—that stars live forever. In fact, even the longest-lived stars will use up all their fuel and burn out after about 100 billion years. This is long before enough light has arrived at Earth to make the sky red.


## DARK MATTER

The Big Bang has enormous explanatory power. Nevertheless, it has serious problems. For one it is difficult to understand where galaxies like our Milky Way came from.

The fireball of the Big Bang was a mix of particles of matter and light. The matter would have affected the light. For instance, if the matter had curdled into clumps, this would be reflected in the afterglow of the Big Bang—it would not be uniform all over the sky today but would be brighter in some places than others. The fact that the afterglow is even all around the sky means that matter in the fireball of the Big Bang must have been spread about extremely smoothly. But we know that it could not be spread completely smoothly. After all, today's Universe is clumpy, with galaxies of stars and clusters of galaxies and great voids of empty space in between. At some point, therefore, the matter in the Universe must have gone from being smoothly distributed throughout space to being clumpy. And the start of this process should be visible in the cosmic background radiation.

Sure enough, in 1992, very slight variations in the brightness of the afterglow of the Big Bang were discovered by NASA's Cosmic Background Explorer Satellite, COBE. These cosmic ripples—one of the scientists involved was more picturesque in likening them to "the face of God"—showed that, about 450,000 years after the Big Bang, some parts of the Universe were a few thousandths of a per cent denser than others. Somehow, these barely noticeable clumps of matter—the "seeds" of structure—had to grow to form the great clusters of galaxies we see in today's Universe. But there is a problem.

Clumps of matter grow to become bigger clumps because of gravity. Basically, if a region has slightly more matter than a neighbouring region, its stronger gravity will ensure that it will steal yet more matter from its neighbour. Just as the richer get richer and the poor get poorer, the denser regions of the Universe will get ever denser until, eventually, they become the galaxies we see around us today. The problem the theorists noticed was that 13.7 billion years was not enough time for gravity to make today's galaxies out of the tiny clumps of matter seen by the COBE satellite. The only way they could do it was if there was much more matter in the Universe than was tied up in visible stars.

Actually, there was strong evidence for missing matter close to home. Spiral galaxies like our own Milky Way are like giant whirlpools of stars, only their stars turn out to be whirling about their centres far too fast. By rights, they should fly off into intergalactic space just as you would be flung off a merry-go-round that someone had spun too fast. The extraordinary explanation that the world's astronomers have come up with is that galaxies like our Milky Way actually contain about 10 times as much matter as is visible in stars. They call the invisible matter dark matter. Nobody knows what it is. However, the extra gravity of the dark matter holds the stars in their orbits and stops them from flying off into intergalactic space.

If the Universe as a whole contains 10 times as much dark matter as ordinary matter, the extra gravity is just enough to turn the clumps of matter seen by COBE into today's galaxy clusters in the 13.7 billion years since the Universe was born. The Big Bang picture is saved.[4] The price is the addition of a lot of dark matter, whose identity nobody knows—well, almost, nobody. In the words of Douglas Adams in *Mostly Harmless:* "For a long period of time there was much speculation and controversy about where the so-called 'missing matter' of the Universe had gotten to. All over the Galaxy the science departments of all the major universities were acquiring more and elaborate equipment to probe and search the hearts of distant galaxies, and then the very centre and the very edges of the whole Universe, but when eventually it was tracked down it turned out in fact to be all the stuff which the equipment had been packed in!"

## INFLATION

The fact that the standard Big Bang picture does not provide enough time for matter to clump into galaxies is not the only problem with the scenario. There is another, arguably more serious, one. It concerns the smoothness of the cosmic background radiation.

Things reach the same temperature when heat travels from a hot body to a cold body. For instance, if you put your cold hand on a hot water bottle, heat will flow from the bottle until your hand

---

[4]Actually, there is thought to be between 6 and 7 times as much dark matter as ordinary matter. This is because the stars account for only about half the ordinary matter. The rest, which may be in the form of dim gas clouds between the galaxies, has not yet been identified.

reaches the same temperature. The cosmic background radiation is basically all at the same temperature. This means that, as the early Universe grew in size, and some bits lagged behind others in temperature, heat always flowed into them from a warmer bit, equalising the temperature.

The problem arises if you imagine the expansion of the Universe running backwards like a movie in reverse. At the time that the cosmic background radiation last had any contact with matter— about 450,000 years after the Big Bang—bits of the Universe that today are on opposite sides of the sky were too far apart for heat to flow from one to the other. The maximum speed it could flow is the speed of light, and the 450,000 years the Universe had been in-existence was simply not long enough. So how is it that the cosmic background radiation is the same temperature everywhere today?

Physicists have come up with an extraordinary answer. Heat could have flowed back and forth throughout the Universe, equalising the temperature, only if the early Universe was much smaller than our backward-running movie would imply. If regions were much closer together than expected, there would have been plenty of time for heat to flow from hot to cold regions and equalise the temperature. But if the Universe was much smaller earlier on, it must have put on a big spurt of growth to get to its present size.

According to the theory of inflation, the Universe "inflated" during its first split-second of existence, undergoing a phenomenally violent expansion. What drove the expansion was a peculiar property of the vacuum of empty space, although that's still hazy to physicists. The point is that there was this enormously fast expansion, which very quickly ran out of steam, and then the more sedate expansion that we see today took over. If the normal Big Bang expansion is likened to the explosion of a stick of dynamite, inflation can be likened to a nuclear explosion. "The standard Big Bang theory says nothing about what banged, why it banged or what happened before it banged," says inflation pioneer Alan Guth. Inflation is at least an attempt to address such questions.

With inflation plus dark matter tagged on, the Big Bang scenario can be rescued. In fact, when astronomers talk of the Big Bang these days, they often mean the Big Bang plus inflation plus dark matter. However, inflation and dark matter are not as well-founded ideas as the Big Bang. Beyond any doubt, we know that the Universe began in a hot dense state and has been expanding and cooling ever since—the Big Bang scenario. That inflation happened is still not certain, and nobody has yet discovered the identity of dark matter.

One of the pluses of inflation is that it provides a possible explanation of the origins of structures such as galaxies in today's Universe. For such structures to have formed, there must have been some kind of unevenness in the Universe at a very early stage. That primordial roughness could have been caused by so-called quantum fluctuations. Basically, the laws of microscopic physics cause extremely small regions of space and matter to jiggle about restlessly like water in a boiling saucepan. Such fluctuations in the density of matter were minuscule—smaller even than present-day atoms. However, the phenomenal expansion of space caused by inflation would actually have enhanced them, blowing them up to noticeable size. Bizarrely, the largest structures in today's Universe—great clusters of galaxies—may have been spawned by "seeds" smaller than atoms!

Inflation, however, predicts something about our Universe that does not seem to accord with the facts. Currently, the Universe is expanding. However, the gravity of all the matter in the Universe is acting to brake the expansion. There are two main possibilities. One is that the Universe contains sufficient matter to eventually slow and reverse its expansion, causing the Universe to collapse back down to a Big Crunch, a sort of mirror image of the Big Bang in which the Universe was born. The other is that it contains insufficient matter and goes on expanding forever. Inflation predicts that the Universe should be balanced on the knife edge between these two possibilities. It will continue expanding, but slowing down all the time, and finally running out of steam only in the infinite future. For this to happen, the Universe must have what is known as the critical mass. The problem is that, even when all the matter in the Universe—visible matter and dark matter—is added up, it amounts to only about a third of the critical mass. Inflation, it would seem, is a nonstarter. Well, that's how it seemed—until a sensational discovery was made in 1998.

## DARK ENERGY

Two teams were observing "supernovas"—exploding stars—in distant galaxies. One team was led by American Saul Perlmutter and the other by Australians Nick Suntzeff and Brian Schmidt. Supernovas are exploding stars that often outshine their parent galaxy and so can be seen at great distances out into the Universe. The kind the two teams of astronomers were looking at were known as Type Ia supernovas. They have the property that, when they detonate, they always shine with the same peak luminosity. So if you see one that is fainter than another, you know it is farther away.

What the astronomers saw, however, was that the ones that were farther away were fainter than they ought to be, taking into account their distance from Earth. The only way to explain what they were seeing was that the Universe's expansion had speeded up since the stars exploded, pushing them farther away than expected and making them appear fainter.

It was a bombshell dropped into the world of science. The sole force affecting the galaxies ought to be their mutual gravitational pull. That should be braking the expansion, not speeding it up.

The only thing that could be accelerating things was space itself. Contrary to all expectations, it could not be empty. It must contain some kind of weird stuff unknown to science—"dark energy"—that was exerting a kind of cosmic repulsion on the Universe, countering gravity and driving the galaxies apart.

Physicists are totally at sea when it comes to understanding dark energy. Their best theory—quantum mechanics—predicts an energy associated with empty space that is 1 followed by 123 zeroes bigger than Perlmutter observed! Nobel laureate Steven Weinberg has described this as "the worst failure of an order-of-magnitude estimate in the history of science."

Despite this embarrassment, the dark energy has at least one desirable consequence. Recall that inflation requires the Universe to have the critical mass but that all the matter in the Universe adds up to only about a third of the critical mass. Well all forms of energy, as Einstein discovered, have an effective mass. And that includes the dark energy. In fact, it turns out to account for about two-thirds of the critical mass, so that the Universe has exactly the critical mass—just what is predicted by inflation.

Although nobody knows what the dark energy is, one possibility is that it is associated with the repulsive force of empty space proposed by Einstein. In science, it seems, all things begin and end with Einstein. His biggest mistake may yet turn out to be his biggest success.

It is worth stressing, however, that the Big Bang, for all its successes, is still basically a description of how our Universe has evolved from a superdense, superhot state to its present state, with galaxies, stars, and planets. How it all began is still shrouded in mystery.

## TO THE SINGULARITY AND BEYOND

Imagine the expansion of the Universe running backwards again like a movie in reverse. As the Universe shrinks down to a speck, its matter content becomes ever more compressed and ever hotter. In fact, there is no limit to this process. At the instant the Universe's expansion began—the moment of its birth—it was infinitely dense and infinitely hot. Physicists call the point when something skyrockets to infinity a singularity. According to the standard Big Bang picture, the Universe was therefore born in a singularity.

The other place where Einstein's theory of gravity predicts a singularity is at the heart of a black hole. In this case the matter of a catastrophically shrinking star eventually becomes compressed into zero volume and therefore becomes infinitely dense and infinitely hot.

"Black holes," as someone once said, "are where God divided by zero."[5]

A singularity is a nonsense. When such a monstrous entity pops up in a theory of physics, it is telling us that the theory—in this case, Einstein's theory of gravity—is faulty. We are stretching it beyond the domain where it has anything sensible to say about the world. This is not surprising. General relativity is a theory of the very large. In its earliest stages, however, the Universe was smaller than an atom. And the theory of the atomic realm is quantum theory.

Normally, there is no overlap between these two towering monuments of 20th-century physics. However, they come into conflict at the heart of black holes and at the birth of the Universe. If we are ever going to understand how the Universe came into being, we are going to have to find a better description of reality than Einstein's theory of gravity. We need a quantum theory of gravity.

The task of finding such a theory is formidable because of the fundamental incompatibility between general relativity and quantum theory. General relativity, like every theory of physics before it, is a recipe for predicting the future. If a planet is here now, in a day's time it will have moved over there, by following this path. All these things are predictable with 100 per cent certainty. Quantum theory, however, is a recipe for predicting probabilities. If an atom is flying through space, all we can predict is its probable final position, its probable path. Quantum theory therefore undermines the very foundation stones of general relativity.

Currently, physicists are trying to discover the elusive quantum theory of gravity by a number of routes. Undoubtedly, the one getting the most publicity is superstring theory, which views the fundamental building blocks of matter not as pointlike particles but as

---

[5]Actually, there is a subtle distinction between the singularities at the heart of a black hole and the Big Bang. The former is a singularity in time and the latter a singularity in space.

ultra-tiny pieces of "string." The string—superconcentrated mass-energy—can vibrate just like a violin string, and each distinct vibration "mode" corresponds to a fundamental particle such as an electron or a photon.

What excites string theorists is that some form of gravity—although not necessarily general relativity—is automatically contained within string theory. One slight complication is that the strings of string theory vibrate in a 10-dimensional world, which means there have to exist an additional six space dimensions too small for us to have noticed. Another problem is that string theory involves such horrendously complicated mathematics that it has so far proved impossible to make a prediction with it that can be tested against reality.

No one knows how close or how far away we are to possessing a quantum theory of gravity. But without it there is no hope of travelling those last tantalising steps back to the beginning of the Universe. However, some of the things that must happen along the route are clear.

Think of the expansion of the Universe in reverse again. At first the Universe will shrink at the same rate in all directions. This is because the Universe is pretty much the same in all directions. But *pretty much the same* is not the same as *exactly the same*. Undoubtedly, there will be slightly more galaxies in one direction than another. In the early stages of the contraction this imbalance will have no noticeable effect. However, as the Universe shrinks down to zero volume, such matter irregularities will become ever more magnified. So when the body shrinks to zero volume, the final stages of the collapse will be wildly chaotic. Gravity—warped space-time—will vary wildly depending on the direction from which the singularity is approached by an in-falling body.

Very close to the singularity, the warpage of space-time will become so violent and chaotic that space and time will actually shatter, splitting into myriad droplets. Concepts like "before" and "after" now lose all meaning. So too do concepts like "distance" and "direction." An impenetrable fog blocks the view ahead. It shrouds the mysterious domain of quantum gravity, where no theory yet exists to act as our guide.

But deep in that fog lie the answers to science's most pressing questions. Where did the Universe come from? Why did it burst into being in a Big Bang 13.7 billion years ago? What, if anything, existed before the Big Bang?

The fervent hope is that, when at last we manage to mesh together our theory of the very small with our theory of the very large, we will find the answers to these questions. Then we will come face to face with the ultimate question: How could something have come from nothing? "It is enough to hold a stone in your hand," wrote Jostein Gaarder in *Sophie's World*. "The universe would have been equally incomprehensible if it had only consisted of that one stone the size of an orange. The question would be just as impenetrable: Where did this stone come from?"

# GLOSSARY

ABSOLUTE ZERO Lowest temperature attainable. As a body is cooled, its atoms move more and more sluggishly. At absolute zero, equivalent to −273.15 on the Celsius scale, they cease to move altogether. (Actually, this is not entirely true since the Heisenberg uncertainty principle produces a residual jitter even at absolute zero.)

ACCRETION DISC CD-shaped disc of in-swirling matter that forms around a strong source of gravity such as a black hole. Since gravity weakens with distance from its source, matter in the outer portion of the disc orbits more slowly than in the inner portion. Friction between regions where matter is travelling at different speeds heats the disc to millions of degrees. Quasars are thought to owe their prodigious brightness to ferociously hot accretion discs surrounding "supermassive" black holes.

ALPHA CENTAURI The nearest star system to the Sun. It consists of three stars and is 4.3 light-years distant.

ALPHA DECAY The spitting out of a high-speed alpha particle by a large, unstable nucleus in an attempt to turn itself into a lighter, stable nucleus.

ALPHA PARTICLE A bound state of two protons and two neutrons—essentially a helium nucleus—that rockets out of an unstable
nucleus during radioactive alpha decay.

ANTHROPIC PRINCIPLE The idea that the Universe is the way it is
because, if it was not, we would not be here to notice it. In other
words, the fact of our existence is an important scientific observation.

ANTIMATTER Term for a large accumulation of antiparticles. Anti-
protons, antineutrons, and positrons can in fact come together to
make anti-atoms. And there is nothing in principle to rule out the
possibility of antistars, antiplanets, or antilife. One of the greatest
mysteries of physics is why we appear to live in a Universe made solely
of matter when the laws of physics seem to predict a pretty much 50/
50 mix of matter and antimatter.

ANTIPARTICLE Every subatomic particle has an associated antiparticle with opposite properties, such as electrical charge. For instance,
the negatively charged electron is twinned with a positively charged
antiparticle known as the positron. When a particle and its antiparticle meet, they self-destruct, or "annihilate," in a flash of high-energy
light, or gamma rays.

ATOM The building block of all normal matter. An atom consists of
a nucleus orbited by a cloud of electrons. The positive charge of the
nucleus is exactly balanced by the negative charge of the electrons. An
atom is about one 10-millionth of a millimetre across.

ATOMIC ENERGY See Nuclear Energy.

ATOMIC NUCLEUS The tight cluster of protons and neutrons (a
single proton in the case of hydrogen) at the centre of an atom. The
nucleus contains more than 99.9 per cent of the mass of an atom.

BIG BANG  The titanic explosion in which the Universe is thought to have been born 13.7 billion years ago. "Explosion" is actually a misnomer since the Big Bang happened everywhere at once and there was no preexisting void into which the Universe erupted. Space, time, and energy all came into being in the Big Bang.

BIG BANG THEORY  The idea that the Universe began in a super-dense, superhot state 13.7 billion years ago and has been expanding and cooling ever since.

BIG CRUNCH  If there is enough matter in the Universe, its gravity will one day halt and reverse the Universe's expansion so that it shrinks down to a Big Crunch. This is a sort of mirror image of the Big Bang.

BLACK BODY  A body that absorbs all the heat that falls on it. The heat is shared among the atoms in such a way that the heat radiation it gives out takes no account of what the body is made of but depends solely on its temperature and has a characteristic and easily recognisable form. The stars are approximate black bodies.

BLACK HOLE  The grossly warped space-time left behind when a massive body's gravity causes it to shrink down to a point. Nothing, not even light, can escape—hence a black hole's blackness. The Universe appears to contain at least two distinct types of black hole—stellar-sized black holes that form when very massive stars can no longer generate internal heat to counterbalance the gravity trying to crush them and "supermassive" black holes. Most galaxies appear to have a supermassive black hole in their heart. They range from millions of times the mass of the Sun in our Milky Way to billions of solar masses in the powerful quasars.

BOSE-EINSTEIN CONDENSATION  Phenomenon in which all the microscopic particles in a body suddenly crowd into the same state.

The particles must be bosons and the temperature must generally be very low. Helium atoms, for instance, crowd into the same state below –271 degrees Celsius, turning liquid helium into a superfluid.

BOSON A microscopic particle with integer spin—that is, 0 units, 1 unit, 2 units, and so on. By virtue of their spin, such particles are hugely gregarious, participating in collective behaviour that leads to lasers, superfluids, and superconductors.

BOYLE'S LAW The observation that the volume of a gas is inversely proportional to its pressure—that is, doubling the pressure halves the volume.

BROWNIAN MOTION The random, jittery motion of a large body under machine-gun bombardment from smaller bodies. The most famous instance is of pollen grains zigzagging through water as they are repeatedly hit by water molecules. The phenomenon, discovered by botanist Robert Brown in 1827 and triumphantly explained by Einstein in 1905, was powerful proof of the existence of atoms.

CAUSALITY The idea that a cause always precedes an effect. Causality is a much-cherished principle in physics. However, quantum events such as the decay of atoms appear to be effects with no prior cause.

CHANDRASEKHAR LIMIT The largest possible mass for a white dwarf. It depends on a star's chemical composition, but for a white dwarf made of helium it is about 44 per cent more massive than the Sun. For a star bigger than this, the electron degeneracy pressure inside is insufficient to prevent gravity from crushing the star farther.

CHARGE-COUPLED DEVICE (CCD) Supersensitive electronic light detector that can register close to 100 per cent of the light that falls on it. Since photographic plates register a mere 1 per cent, CCDs allow a telescope to perform as well as a telescope with 100 times the light-collecting area.

CHEMICAL BOND  The "glue" that sticks atoms together to make molecules.

CHRONOLOGY PROTECTION CONJECTURE  The stricture that time travel is impossible. No one has yet managed to prove it—in fact, the laws of physics appear to permit time travel—but physicists such as Stephen Hawking remain convinced that some, as-yet-undiscovered law of nature forbids time machines.

CLASSICAL PHYSICS  Nonquantum physics. In effect, all physics before 1900 when the German physicist Max Planck first proposed that energy might come in discrete chunks, or quanta. Einstein was the first to realise that this idea was totally incompatible with all physics that had gone before.

CLOSED TIME-LIKE CURVE (CTC)  Region of space-time so dramatically warped that time loops back on itself in much the same way that space loops back on itself on an athletics track. A CTC, in common parlance, is a time machine. It is permitted to exist by the current laws of physics.

COMET  Small icy body—usually mere kilometres across—that orbits a star. Most comets orbit the Sun beyond the outermost planets in an enormous cloud known as the Oort Cloud. Like asteroids, comets are builders' rubble left over from the formation of the planets.

COMPTON EFFECT  The recoil of an electron when exposed to high-energy light just as if the electron is a tiny billiard ball struck by another tiny billiard ball. The effect is a graphic demonstration that light is ultimately made of tiny bulletlike particles, or photons.

CONDUCTOR  A material through which an electrical current can flow.

CONSERVATION LAW  Law of physics that expresses the fact that a quantity can never change. For instance, the conservation of energy states that energy can never be created or destroyed, only converted from one form to another. For example, the chemical energy of petrol can be converted into the energy of motion of a car.

CONSERVATION OF ENERGY  Principle that energy can never be created or destroyed, only converted from one form to another.

COOPER PAIR  Two electrons with opposite spin that pair up in some metals at extremely low temperature. Cooper pairs, unlike individual electrons, are bosons. Consequently, they can crowd into the same state, moving together in lockstep through the metal like an irresistible army on the move. The electrical current in such a "superconductor" can run forever.

COPERNICAN PRINCIPLE  The idea that there is nothing special about our position in the Universe, in either space or time. This is a generalised version of Copernicus's recognition that Earth is not in a special position at the centre of the solar system but is just another planet circling the Sun.

COSMIC BACKGROUND RADIATION  The "afterglow" of the Big Bang fireball. Incredibly, it still permeates all of space 13.7 billion years after the event, a tepid radiation corresponding to a temperature of −270 degrees Celsius.

COSMIC RAYS  High-speed atomic nuclei, mostly protons, from space. Low-energy ones come from the Sun; high-energy ones probably come from supernovas. The origin of ultra-high-energy cosmic rays, particles millions of times more energetic than anything we can currently produce on Earth, is one of the great unsolved puzzles of astronomy.

COSMOLOGY  The ultimate science. The science whose subject matter is the origin, evolution, and fate of the entire Universe.

COSMOS  Another word for Universe.

DARK ENERGY  Mysterious "material" with repulsive gravity. Discovered unexpectedly in 1998, it is invisible, fills all of space and appears to be pushing apart the galaxies and speeding up the expansion of the Universe. Nobody has much of a clue what it is.

DARK MATTER  Matter in the Universe that gives out no light. Astronomers know it exists because the gravity of the invisible stuff bends the paths of visible stars and galaxies as they fly through space. There is between 6 and 7 times as much dark matter in the Universe as ordinary, light-emitting matter. The identity of the dark matter is the outstanding problem of astronomy.

DECOHERENCE  The mechanism that destroys the weird quantum nature of a body—so that, for instance, it appears localised rather than in many different places simultaneously. Decoherence occurs if the outside world gets to "know" about the body. The knowledge may be taken away by a single photon of light or an air molecule that bounces off the body. Since big bodies like tables are continually struck by photons and air molecules and cannot remain isolated from their surroundings for long, they lose their ability to be in many places at once in a fantastically short time—far too short for us to notice.

DEGENERACY PRESSURE  The bee-in-a-box-like pressure exerted by electrons squeezed into a small volume of space. A consequence of the Heisenberg uncertainty principle, it arises because a microscopic particle whose location is known very well necessarily has a large uncertainty in its velocity. The degeneracy pressure of electrons prevents white dwarfs from shrinking under their own gravity, whereas the degeneracy pressure of neutrons does the same thing for neutron stars.

DENSITY  The mass of an object divided by its volume. Air has a low density, and iron has a high density.

DIMENSION  An independent direction in space-time. The familiar world around us has three space dimensions (east–west, north–south, up-down) and one of time (past-future). Superstring theory requires the Universe to have six extra space dimensions. These differ radically from the other dimensions because they are rolled up very small.

DOUBLE SLIT EXPERIMENT  Experiment in which microscopic particles are shot at a screen with two closely spaced, parallel slits cut in it. On the far side of the screen, the particles mingle, or "interfere," with each other to produce a characteristic "interference pattern" on a second screen. The bizarre thing is that the pattern forms even if the particles are shot at the slits one at a time, with long gaps between— in other words, when there is no possibility of them mingling with each other. This result, claimed Richard Feynman, highlighted the "central mystery" of quantum theory.

ELECTRIC CHARGE  A property of microscopic particles that comes in two types—positive and negative. Electrons, for instance, carry a negative charge and protons a positive charge. Particles with the same charge repel each other, while particles with unlike charge attract.

ELECTRIC CURRENT  A river of charged particles, usually electrons, that can flow through a conductor.

ELECTRIC FIELD  The field of force that surrounds an electric charge.

ELECTROMAGNETIC FORCE  One of the four fundamental forces of nature. It is responsible for gluing together all ordinary matter, including the atoms in our bodies and the atoms in the rocks beneath our feet.

ELECTROMAGNETIC WAVE  A wave that consists of an electric field that periodically grows and dies, alternating with a magnetic field that periodically dies and grows. An electromagnetic wave is generated by a vibrating electric charge and travels through space at the speed of light.

ELECTRON  Negatively charged subatomic particle typically found orbiting the nucleus of an atom. As far as anyone can tell, it is a truly elementary particle, incapable of being subdivided.

ELEMENT  A substance that cannot be reduced any further by chemical means. All atoms of a given element possess the same number of protons in their nucleus. For instance, all atoms of hydrogen have one proton, all atoms of chlorine have 17, and so on.

ENERGY  A quantity that is almost impossible to define! Energy can never be created or destroyed, only converted from one form to another. Among the many familiar forms are heat energy, energy of motion, electrical energy, and sound energy.

ENTANGLEMENT  The intermingling of two or more microscopic particles so that they lose their individuality and in many ways behave as a single entity.

EVENT HORIZON  The one-way "membrane" that surrounds a black hole. Anything that falls through—whether matter or light—can never get out again.

EXOTIC MATTER  Hypothetical matter with repulsive gravity.

EXPANDING UNIVERSE  The fleeing of the galaxies from each other in the aftermath of the Big Bang.

FERMION  A microscopic particle with half-integer spin—that is, $\frac{1}{2}$ unit, $\frac{3}{2}$ units, $\frac{5}{2}$ units, and so on. By virtue of their spin, such particles shun each other. Their unsociability is the reason that atoms exist and the ground beneath our feet is solid.

FRAME DRAGGING  The dragging around of space-time by a massive rotating body. The effect is very small—though potentially measurable—in the vicinity of Earth but enormous near a fast-rotating black hole. Such a black hole sits at the eye of a tornado of whirling space-time.

FUNDAMENTAL FORCE  One of the four basic forces that are believed to underlie all phenomena. The four forces are the gravitational force, electromagnetic force, strong force, and weak force. The strong suspicion among physicists is that these forces are actually facets of a single superforce. In fact, experiments have already shown the electromagnetic and weak forces to be different sides of the same coin.

FUNDAMENTAL PARTICLE  One of the basic building blocks of all matter. Currently, physicists believe there are six different quarks and six different leptons, making a total of 12 truly fundamental particles. The hope is that the quarks will turn out to be merely different faces of the leptons.

FUSION  See Nuclear Fusion.

GALAXY  One of the basic building blocks of the Universe. Galaxies are great islands of stars. Our own island, the Milky Way, is spiral in shape and contains about 200,000 million stars.

GAS  Collection of atoms that fly about through space like a swarm of tiny bees.

GENERAL THEORY OF RELATIVITY  Einstein's theory of gravity that shows gravity to be nothing more than the warpage of space-time. The theory incorporates several ideas that were not incorporated in Newton's theory of gravity. One was that nothing, not even gravity, can travel faster than light. Another was that all forms of energy have mass and so are sources of gravity. Among other things, the theory predicted black holes, the expanding Universe, and that gravity would bend the path of light.

GEODESIC  The shortest path between two points in warped, or curved, space.

GRAVITATIONAL FORCE  The weakest of the four fundamental forces of nature. Gravity is approximately described by Newton's universal law of gravity but more accurately described by Einstein's theory of gravity—the general theory of relativity. General relativity breaks down at the singularity at the heart of a black hole and the singularity at the birth of the Universe. Physicists are currently looking for a better description of gravity. The theory, already dubbed quantum gravity, will explain gravity in terms of the exchange of particles called gravitons.

GRAVITATIONAL LIGHT BENDING  The bending of the trajectory of light that passes by a massive body. Because the space in the vicinity of such a body is warped like a valley, the light has no choice but to travel along a curved path.

GRAVITATIONAL RED SHIFT  The loss of energy as light climbs out of the valley in space-time around a massive celestial body. Since the "colour" of light is related to its energy, with red light having less energy than blue light, astronomers talk of light being shifted to the red end of the spectrum or "red-shifted."

GRAVITATIONAL WAVE  A ripple spreading out through space-time. Gravitational waves are generated by violent motions of mass, such as the merger of black holes. Because they are weak, they have not yet been detected directly.

GRAVITY  See Gravitational Force.

HALF-LIFE  The time it takes half the nuclei in a radioactive sample to disintegrate. After one half-life, half the atoms will be left; after two half-lives, a quarter; after three, an eighth, and so on. Half-lives can vary from the merest split-second to many billions of years.

HEISENBERG UNCERTAINTY PRINCIPLE  A principle of quantum theory that there are pairs of quantities such as a particle's location and speed that cannot simultaneously be known with absolute precision. The uncertainty principle puts a limit on how well the product of such a pair of quantities can be known. In practice, this means that if the speed of a particle is known precisely, it is impossible to have any idea where the particle is. Conversely, if the location is known with certainty, the particle's speed is unknown. By limiting what we can know, the Heisenberg uncertainty principle imposes "fuzziness" on nature. If we look too closely, everything blurs like a newspaper picture dissolving into meaningless dots.

HELIUM  Second lightest element in nature and the only one to have been discovered on the Sun before it was discovered on Earth. Helium is the second most common element in the Universe after hydrogen, accounting for about 10 per cent of all atoms.

HORIZON  The Universe has a horizon much like the horizon that surrounds a ship at sea. The reason for the Universe's horizon is that light has a finite speed and the Universe has been in existence for only a finite time. This means that we only see objects whose light has had time to reach us since the Big Bang. The observable universe is therefore like a bubble centred on Earth, with the horizon being the surface of the bubble. Every day the Universe gets older (by one day), so every day the horizon expands outwards and new things become visible, just like ships coming over the horizon at sea.

HORIZON PROBLEM  The problem that far-flung parts of the Universe that could never have been in contact with each other, even in the Big Bang, have almost identical properties such as density and temperature. Technically, they were always beyond each other's horizon. The theory of inflation provides a way for such regions to have been in contact in the Big Bang and so can potentially solve the horizon problem.

HYDROGEN  The lightest element in nature. A hydrogen atom consists of a single proton orbited by a single electron. Close to 90 per cent of all atoms in the Universe are hydrogen atoms.

HYDROGEN BURNING  The fusion of hydrogen into helium accompanied by the liberation of large quantities of nuclear binding energy. This is the power source of the Sun and most stars.

HYDROSTATIC EQUILIBRIUM  The state in which the gravitational force trying to crush a star is perfectly balanced by the force of its hot gas pushing outwards.

INERTIA The tendency for a massive body, once set in motion, to keep on moving, at constant speed in a straight line in unwarped space and along a geodesic in warped space. Nobody knows the origin of inertia.

INERTIAL FORCE A force we invent to explain a motion that is actually due to nothing more than inertia. A good example is centrifugal force. There is no such force flinging us outwards in a car rounding a sharp corner. We are simply continuing to move in a straight line because of our inertia, and the interior of the car, because it is moving along a curved path, intercepts us.

INFLATION, THEORY OF Idea that in the first split-second of its creation the Universe underwent a fantastically fast expansion. In a sense, inflation preceded the conventional Big Bang explosion. If the Big Bang is likened to the explosion of a grenade, inflation was like the explosion of an H-bomb. Inflation can solve some problems with the Big Bang theory such as the horizon problem.

INFRARED Type of invisible light that is given out by warm bodies.

INTERFERENCE The ability of two waves passing through each other to mingle, reinforcing where their peaks coincide and cancelling where the peaks of one coincide with the troughs of another.

INTERFERENCE PATTERN Pattern of light and dark stripes that appears on a screen illuminated by light from two sources. The pattern is due to the light from the two sources reinforcing at some places on the screen and cancelling at other places.

INTERSTELLAR MEDIUM The tenuous gas and dust floating between the stars. In the vicinity of the Sun this gas comprises about one hydrogen atom in every 3 cubic centimetres, making it a vacuum far better than anything achievable on Earth.

INTERSTELLAR SPACE  The space between the stars.

ION  An atom or molecule that has been stripped of one or more of its orbiting electrons and so has a net positive electrical charge.

ISOTOPE  One possible form of an element. Isotopes are distinguishable by their differing masses. For instance, chlorine comes in two stable isotopes, with a mass of 35 and 37. The mass difference is due to a differing number of neutrons in their nuclei. For instance, chlorine-35 contains 18 neutrons and chlorine-37 contains 20 neutrons. (Both contain the same number of protons—17—since this determines the identity of an element.)

JOULE  The standard scientific unit of energy. The energy of motion of a flying cricket ball is about 10 joules; the chemical energy provided by a single slice of bread is about 100,000 joules; and the electrical energy of a lightning discharge is about 10 billion joules.

LAMBDA POINT  Temperature below which liquid helium begins to turn into a superfluid.

LASER  Light source in which the gregarious nature of photons—bosons—comes to the fore. Specifically, the more photons there are passing through a material the greater the probability that other atoms will emit others with the same properties. The result is an avalanche of photons all travelling in lockstep.

LIGHT, CONSTANCY OF  The peculiarity that in our Universe the speed of light in empty space is always the same, irrespective of the speed of the source of light or of anyone observing the light. This is one of two cornerstones of Einstein's special theory of relativity, the other being the principle of relativity.

LIGHT, SPEED OF   The cosmic speed limit—300,000 kilometres per second.

LIGHT BENDING   See Gravitational Light Bending.

LIGHT-YEAR   Convenient unit for expressing distances in the Universe. It is simply the distance that light travels in one year in a vacuum, which turns out to be 9.46 trillion kilometres.

LORENTZ CONTRACTION   The contraction of a body moving relative to an "observer." The observer sees the body shrink in the direction of its motion. The effect is noticeable only when the body is moving close to the speed of light with respect to the observer.

LUMINOSITY   The total amount of light pumped into space each second by a celestial body such as a star.

MAGNETIC FIELD   The field of force that surrounds a magnet.

MANY WORLDS   The idea that quantum theory describes everything, not simply the microscopic world of atoms and their constituents. Since quantum theory permits an atom to be in two places at once, this must mean that a table can be in two places at once. According to the Many Worlds idea, however, the mind of the person observing the table splits into two—one that perceives the table to be in one place and another that perceives it to be in another. The two minds exist in separate realities, or universes.

MASS   A measure of the amount of matter in a body. Mass is the most concentrated form of energy. A single gram contains the same amount of energy as 100,000 tonnes of dynamite.

MAXWELL'S EQUATIONS OF ELECTROMAGNETISM   The handful of elegant equations, written down by James Clerk Maxwell in

1868, that neatly summarise all electrical and magnetic phenomena. The equations reveal that light is an electromagnetic wave.

MILKY WAY  Our galaxy.

MOLECULE  Collection of atoms glued together by electromagnetic forces. One atom, carbon, can link with itself and other atoms to make a huge number of molecules. For this reason, chemists divide molecules into "organic"—those based on carbon—and "inorganic"—the rest.

MOMENTUM  The momentum of a body is a measure of how much effort is required to stop it. For instance, an oil tanker, even though it may be going at only a few kilometres an hour, is far harder to stop than a Formula 1 racing car going 200 kilometres per hour. The oil tanker is said to have more momentum.

MOMENTUM, CONSERVATION OF  Principle that momentum can never be created or destroyed.

MULTIVERSE  Hypothetical enlargement of the cosmos in which our Universe turns out to be one among an enormous number of separate and distinct universes. Most universes are dead and uninteresting. Only in a tiny subset do the laws of physics promote the emergence of stars, planets, and life.

MUON  Short-lived subatomic particle that behaves like a heavy version of the electron.

NEUTRINO  Neutral subatomic particle with a very small mass that travels very close to the speed of light. Neutrinos, of which there are three kinds, hardly ever interact with matter. However, when created in huge numbers, they can blow a star apart in a supernova.

NEUTRON One of the two main building blocks of the atomic nucleus at the centre of atoms. Neutrons have essentially the same mass as protons but carry no electrical charge. They are unstable outside of a nucleus and disintegrate in about 10 minutes.

NEUTRON STAR A star that has shrunk under its own gravity to such an extent that most of its material has been compressed into neutrons. Typically, such a star is only 20 to 30 kilometres across. A sugar cube of neutron star stuff would weigh as much as the entire human race.

NEWTON'S UNIVERSAL LAW OF GRAVITY The idea that all bodies pull on each other across space with a force that depends on the product of their individual masses and the inverse square of their distance apart. In other words, if the distance between the bodies is doubled, the force becomes four times weaker; if it is tripled, nine times weaker; and so on. Newton's theory of gravity is perfectly good for everyday applications but turns out to be an approximation. Einstein provided an improvement in the general theory of relativity.

NONLOCALITY The spooky ability of objects subject to quantum theory to continue to "know" about each other's state even when separated by a large distance.

NUCLEAR ENERGY The excess energy released when one atomic nucleus changes into another atomic nucleus.

NUCLEAR FUSION The welding together of two light nuclei to make a heavier nucleus, a process that results in the liberation of nuclear binding energy. The most important fusion process for human beings is the gluing together of hydrogen nuclei to make helium in the core of the Sun since its by-product is sunlight.

NUCLEAR REACTION  Any process that converts one type of atomic nucleus into another type of atomic nucleus.

NUCLEON  Umbrella term used for protons and neutrons, the two building blocks of the atomic nucleus.

NUCLEUS  See Atomic Nucleus.

PARTICLE ACCELERATOR  Giant machine, often in the shape of a circular racetrack, in which subatomic particles are accelerated to high speed and smashed into each other. In such collisions the energy of motion of the particles becomes available to create new particles.

PARTICLE PHYSICS  The quest to discover the fundamental building blocks and fundamental forces of nature.

PAULI EXCLUSION PRINCIPLE  The prohibition on two microscopic particles (fermions) sharing the same quantum state. The Pauli exclusion stops electrons, which are fermions, from piling on top of each other and, consequently, explains the existence of different atoms and of the variety of the world around us.

PHOTOCELL  A practical device that exploits the photoelectric effect. The interruption of an electric current when a body breaks the light beam falling on a metal is used to control something—for instance, an automatic door at the entrance to a supermarket.

PHOTOELECTRIC EFFECT  The ejection of electrons from the surface of a metal by photons striking the metal.

PHOTON  Particle of light.

PHYSICS, LAWS OF  The fundamental laws that orchestrate the behavior of the Universe.

PLANCK ENERGY  The superhigh energy at which gravity becomes comparable in strength to the other fundamental forces of nature.

PLANCK LENGTH  The fantastically tiny length scale at which gravity becomes comparable in strength to the other fundamental forces of nature. The Planck length is a trillion trillion times smaller than an atom. It corresponds to the Planck energy. Small distances are synonymous with high energies because of the wave nature of matter.

PLASMA  An electrically charged gas of ions and electrons.

POSITRON  Antiparticle of the electron.

PRECESSION OF THE PERIHELION OF MERCURY  The fact that the orbit of Mercury, the planet closest to the Sun, does not follow a straightforward elliptical orbit but rather an elliptical orbit whose nearest point to the Sun gradually moves around the Sun, resulting in the planet tracing out a rosettelike pattern. The explanation is that the gravity of the Sun weakens with distance from the Sun more slowly than in the case of Newtonian gravity, which uniquely predicts elliptical orbits. It weakens more slowly because, in the Einsteinian picture, gravity itself is a source of more gravity.

PRINCIPLE OF EQUIVALENCE  The idea that gravity and acceleration are indistinguishable.

PROTON  One of the two main building blocks of the nucleus. Protons carry a positive electrical charge, equal and opposite to that of electrons.

PULSAR  A rapidly rotating neutron star that sweeps an intense beam of radio waves around the sky much like a lighthouse.

QED  See Quantum Electrodynamics.

QUANTUM  The smallest chunk into which something can be divided. Photons, for instance, are quanta of the electromagnetic field.

QUANTUM COMPUTER  A machine that exploits the fact that quantum systems such as atoms can be in many different states at once to carry out many calculations at once. The best quantum computers can manipulate only a handful of binary digits, or bits, but in principle such computers could massively outperform conventional computers.

QUANTUM ELECTRODYNAMICS  Theory of how light interacts with matter. The theory explains almost everything about the everyday world, from why the ground beneath our feet is solid to how a laser works, from the chemistry of metabolism to the operation of computers.

QUANTUM INDISTINGUISHABILITY  The inability to distinguish between two quantum events. These may be indistinguishable, for instance, because they involve identical particles or simply because the events are not observed. The crucial thing, however, is that the probability waves associated with indistinguishable events interfere. This leads to all manner of quantum phenomena.

QUANTUM NUMBER  A number that specifies a microscopic property that comes in chunks such as the spin or orbital energy of an electron.

QUANTUM PROBABILITY  The chance, or probability, of a microscopic event. Although nature prohibits us from knowing things with certainty, it nevertheless permits us to know the probabilities with certainty.

QUANTUM SUPERPOSITION  Situation in which a quantum object such as an atom is in more than one state at a time. It might, for instance, be in many places simultaneously. It is the interaction, or "interference," between the individual states in the superposition that is the basis of all quantum weirdness. Decoherence prevents such interaction and therefore destroys quantum behaviour.

QUANTUM THEORY  The theory of objects isolated from their surroundings. Because it is very hard to isolate a big object, the theory is essentially a theory of the microscopic world of atoms and their constituents.

QUANTUM TUNNELLING  The apparently miraculous ability of microscopic particles to escape their prisons. For instance, an alpha particle can tunnel through the barrier penning it in the nucleus, the equivalent of a high jumper jumping a 4-metre-high wall. Tunnelling is yet another consequence of the wavelike character of microscopic particles.

QUANTUM UNPREDICTABILITY  The unpredictability of microscopic particles. Their behaviour is unpredictable even in principle. Contrast this with the unpredictability of a coin toss. It is unpredictable only in practice. In principle, if we knew the shape of the coin, the force exerted on it, the air currents around it, and so on, we could predict the outcome.

QUANTUM VACUUM  The quantum picture of empty space. Far from empty, it seethes with ultra-short-lived microscopic particles that are permitted by the Heisenberg uncertainty principle to blink into existence and blink out again.

QUASAR  A galaxy that derives most of its energy from matter heated to millions of degrees as it swirls into a central giant black hole. Quasars can generate as much light as a hundred normal galaxies from a volume smaller than the solar system, making them the most powerful objects in the Universe.

QUBIT A quantum bit, or binary digit. Whereas a normal bit can only represent a "0" or a "1," a qubit can exist in a superposition of the two states, representing a "0" and a "1" simultaneously. Because strings of qubits can represent a large number of numbers simultaneously, they can be used to do a large number of calculations simultaneously.

RADIOACTIVE DECAY The disintegration of unstable heavy atomic nuclei into lighter, stabler atomic nuclei. The process is accompanied by the emission of either alpha particles, beta particles, or gamma rays.

RADIOACTIVITY Property of atoms that undergo radioactive decay.

RADIUM Highly unstable, or radioactive, element discovered by Marie Curie in 1898.

RELATIVITY, GENERAL THEORY OF Einstein's generalisation of his special theory of relativity. General relativity relates what one person sees when looking at another person accelerating relative to them. Because acceleration and gravity are indistinguishable—the principle of equivalence—general relativity is also a theory of gravity.

RELATIVITY, PRINCIPLE OF The observation that all the laws of physics are the same for observers moving at constant speed with respect to each other.

RELATIVITY, SPECIAL THEORY OF Einstein's theory that relates what one person sees when looking at another person moving at constant speed relative to them. It reveals, among other things, that the moving person appears to shrink in the direction of their motion while their time slows down, effects that become ever more marked as they approach the speed of light.

SCANNING TUNNELLING MICROSCOPE (STM) A device that drags an ultrafine needle across the surface of a material and converts the up-and-down motion of the needle into an image of the atomic landscape of the surface.

SCHRÖDINGER EQUATION Equation that governs the way in which the probability wave, or wave function, describing, say a sub-atomic particle, changes with time.

SIMULTANEITY The idea that events that appear to happen at the same time for one person should appear to happen at the same time for everyone in the Universe. Special relativity shows that this idea is mistaken.

SINGULARITY Location where the fabric of space-time ruptures and so cannot be understood by Einstein's theory of gravity, the general theory of relativity. There was a singularity—a point where quantities such as temperature skyrocketed to infinity—at the beginning of the Universe. There is also one in the centre of every black hole.

SOLAR SYSTEM The Sun and its family of planets, moons, comets, and other assorted rubble.

SPACE-TIME In the general theory of relativity, space and time are seen to be essentially the same thing. They are therefore treated as a single entity—space-time. It is the warpage of space-time that turns out to be gravity.

SPECTRAL LINE Atoms and molecules absorb and give out light at characteristic wavelengths. If they swallow more light than they emit, the result is a dark line in the spectrum of a celestial object. Conversely, if they emit more than they swallow, the result is a bright line.

SPECTRUM  The separation of light into its constituent "rainbow" colours.

SPIN  Quantity with no everyday analog. Loosely speaking, subatomic particles with spin behave as if they are tiny spinning tops (only they are not spinning at all!).

STAR  A giant ball of gas that replenishes the heat it loses to space by means of nuclear energy generated in its core.

STRING THEORY  See Superstring Theory.

STRONG NUCLEAR FORCE  The powerful short-range force that holds protons and neutrons together in an atomic nucleus.

SUBATOMIC PARTICLE  A particle smaller than an atom, such as an electron or a neutron.

SUN  The nearest star.

SUPERCONDUCTOR  A material that, when cooled to ultralow temperatures, conducts an electrical current forever—that is, with no resistance. This ability is connected with a change in the conducting particles from fermions to bosons. Specifically, electrons (fermions) pair up to form Cooper pairs (bosons).

SUPERFLUID  A fluid that, below a critical temperature, develops bizarre properties such as the ability to flow uphill and squeeze through impossibly small holes. The best example is liquid helium, which becomes a superfluid below a temperature of 2.17 degrees above absolute zero. Superfluid liquid helium owes its weirdness to quantum theory and the fact that helium atoms are bosons.

SUPERNOVA A cataclysmic explosion of a massive star. A supernova may, for a short time, outshine an entire galaxy of 100 billion ordinary stars. It is thought to leave behind a highly compressed neutron star or even a black hole.

SUPERSTRING THEORY Theory which postulates that the fundamental ingredients of the Universe are tiny strings of matter. The strings vibrate in a space-time of 10 dimensions. The great payoff of this idea is that it may be able to unite, or "unify," quantum theory and the general theory of relativity.

TACHYON Hypothetical particle that lives its life permanently travelling faster than light.

TELEPORTATION The clever use of entanglement to pin down the exact state of a microscopic particle, in apparent violation of what is permitted by the Heisenberg uncertainty principle. This enables the information necessary to reconstruct the state of the particle to be sent to a remote site.

TEMPERATURE The degree of hotness of a body. Related to the energy of motion of the particles that compose it.

THERMODYNAMICS, SECOND LAW OF The decree that entropy, or microscopic disorder of a body, cannot ever decrease. This is equivalent to saying that heat can never flow from a cold to a hot body.

TIME DILATION The slowing down of time for an observer moving close to the speed of light or experiencing strong gravity.

TIME LOOP See Closed Time-Like Curve.

TIME MACHINE See Closed Time-Like Curve.

TIME TRAVEL  Travel into the past or future—in the case of the future, at a rate of more than 1 year per year.

TIME TRAVEL PARADOX  Nonsensical situation that time travel appears to permit. The most famous is the grandfather paradox in which someone goes back in time and shoots their grandfather before he conceives their mother. How then could they have been born to go back in time and commit the act?

TOTAL ECLIPSE OF THE SUN  The coverage of the Sun by the disc of the Moon when the Moon moves between the Sun and Earth.

TWIN PARADOX  The paradox that arises when someone travels at close to light speed to, say, Alpha Centauri and back while their twin stays at home. According to special relativity, the space-travelling twin ages less. However, from another point of view, it is Earth that receded from the space-travelling twin at close to the speed of light and therefore the stay-at-home-twin who ages less. The paradox is resolved by realising that the two situations are not equivalent. The space-travelling twin must undergo a deceleration and an acceleration at the turnaround at Alpha Centauri, and accelerations require general relativity not special relativity.

ULTRAVIOLET  Type of invisible light that is given out by very hot bodies which is responsible for sunburn.

UNCERTAINTY PRINCIPLE  See Heisenberg Uncertainty Principle.

UNIFICATION  The idea that at extremely high energy the four fundamental forces of nature are one, united in a single theoretical framework.

UNIVERSE  All there is. This is a flexible term once used for what we now call the solar system. Later, it was used for what we call the Milky

Way. Now it is used for the sum total of all the galaxies, of which there appear to be about 100 billion within the observable Universe.

UNIVERSE, EXPANSION OF  The fleeing of the galaxies from each other in the aftermath of the Big Bang.

UNIVERSE, OBSERVABLE  All we can see out to the Universe's horizon.

URANIUM  The heaviest naturally occurring element.

VIRTUAL PARTICLE  Subatomic particle that has a fleeting existence, popping into being and popping out again according to the constraint imposed by the Heisenberg uncertainty principle.

VISCOSITY  The internal friction of a liquid. Treacle has high viscosity and water has low viscosity.

WAVE FUNCTION  A mathematical entity that contains all that is knowable about a quantum object such as an atom. The wave function changes in time and space according to the Schrödinger equation.

WAVELENGTH  The distance for a wave to go through a complete oscillation cycle.

WAVE-PARTICLE DUALITY  The ability of a subatomic particle to behave as a localised billiard ball-like particle or a spread-out wave.

WEAK NUCLEAR FORCE  The second force experienced by protons and neutrons in an atomic nucleus, the other being the strong nuclear force. The weak nuclear force can convert a neutron into a proton and so is involved in beta decay.

WHITE DWARF  A star that has run out of fuel and that gravity has compressed until it is about the size of Earth. A white dwarf is supported against further shrinkage by electron degeneracy pressure. A sugar cube of white dwarf material weighs about as much as a family car.

WORMHOLE  A tunnel through space-time that connects widely spaced regions and so provides a shortcut.

X-RAYS  A high-energy form of light.

# Further Reading

## ATOMS AND QUANTUM THEORY

*Quantum: A Guide for the Perplexed*, by Jim Al-Khalili (Weidenfeld & Nicolson, London, 2003).

*Taming the Atom*, by Hans Christian von Baeyer (Penguin, London, 1994).

*Minds, Machines, and the Multiverse*, by Julian Brown (Little Brown, New York, 2000).

*The Magic Furnace*, by Marcus Chown (Vintage, London, 2000).

*The Fabric of Reality*, by David Deutsch (Penguin, London, 1997).

*Thirty Years That Shook Physics*, by George Gamow (Dover, New York, 1985).

*The Great Physicists from Galileo to Einstein*, by George Gamow (Dover, New York, 1988).

*The New Quantum Universe*, by Tony Hey and Patrick Walters, 2nd edition (Cambridge University Press, Cambridge, England, 2004).

*The Feynman Lectures on Physics*, edited by Robert Leighton et al. (Addison-Wesley, New York, 1989).

## RELATIVITY AND COSMOLOGY

*Afterglow of Creation*, by Marcus Chown (University Science Books, Sausalito, California, 1994).

*The Universe Next Door*, by Marcus Chown (Headline, London, 2002).

*Cosmology*, by Edward Harrison (Cambridge University Press, Cambridge, England, 1991).

*The River of Time*, by Igor Novikov (Cambridge University Press, Cambridge, England, 1998).

*Einstein's Legacy*, by Julian Schwinger (Scientific American Library, New York, 1986).

*The Physical Universe*, by Frank Shu (University Science Books, Sausalito, California, 1982).

# INDEX

## A

Acceleration
  curvature of space and, 124–125
  gravity and, 120–122
Adams, Douglas, 32, 151
Aging
  general theory of relativity, 117–118
  gravity effects, 130–131
  special theory of relativity, 93, 95–96
Allen, Woody, 106
Alpha particle
  decay, 9, 38
  definition, 9
  escape from nucleus, 38, 41–42
  scattering studies, 10, 11
  tunnelling, 42–44
Alpher, Ralph, 146
Anaxagoras, 112
Antimatter, 51*n,* 116
Aspect, Alain, 55
Aston, Francis, 111, 112
Atkinson, Robert, 43
Atmosphere, 6–7

Atomic theory
  atomic decay, 9–10, 38
  chemical properties, 75, 77
  duration of atoms, 13–14, 44
  nature of light, 15–16, 18–19
  origins and development, 4–8
  quantum theory and, 14
  size of atoms, 4, 45
  structure and properties of atoms, 9–14, 44, 75–77
  structure and properties of matter, 3–4, 10–11
  types of atoms, 8–9
  uncertainty principle, 44
Atomic weight, 76, 111–112
Attraction, atomic, 10–11

## B

Becquerel, Henri, 9
Beginning of Universe. *See* Big Bang
Bernoulli, Daniel, 4–5
Big Bang
  cosmic background radiation as
    evidence of, 140–141, 145–148,
    151–152

Superposition
  conceptual basis, 27–28
  decoherence, 34
  definition, 25
  interference in, 30, 39
  observability, 31, 32, 33–34, 35
  in quantum computing, 29, 30–31
Superstring theory, 156–157

## T

Tachyons, 94
Teleportation, 57–59
Television static, 148
Temperature
  Big Bang, 145–146
  cosmic background radiation, 147,
    151–152
  electric current in metal, 82, 83
  weight and, 109, 110
Thomson, J. J., 10, 11
Time
  in black holes, 137–138
  concept of past and future, 103–104
  four dimensional space-time, 125–
    126
  general theory of relativity, 130–132
  gravity effects, 130–132
  relationship to space, 101–104
  special theory of relativity, 92–93,
    94–105
  synchronisation, 98–100
  travel in, 94–95, 137$n$
Tunnelling, 38
  current, 7$n$
  in nuclear fusion, 42–44
Twin paradox, 95$n$

## U

Uncertainty principle, 38
  atomic structure and, 44, 47

conceptual basis, 39–42
  empty space and, 50–51
  entanglement and, 57–58
  Pauli exclusion principle and, 73
  physics of stars and, 47–50
  tunnelling and, 43–44
Uranium, 12$n$, 76

## V

Velocity of particle
  exceeding speed of light, 94
  uncertainty principle, 40–42
Vibration
  electron behaviour and properties,
    73, 74
  superstring theory, 156–157
Viscosity, 80
Volume of matter, 3, 45, 49

## W

Wave equation, 24
Wave phenomenon
  ability to penetrate matter, 37–38
  electron properties, 46
  gravity waves, 127–128, 132
  interference in, 29–30
  light as, 17–18, 22–23, 89
  Maxwell's electromagnetic theory,
    88–89
  in quantum theory, 23–24
  in quantum tunnelling, 38
  superposition, 25, 27–28, 39
  vibration, 73–74
Wavelength of light, 18
Weight
  of atoms, 76, 111–112
  of energy, 106–108, 109
  motion and, 110
  temperature and, 109, 110

*Also by Marcus Chown*

ff

# What a Wonderful World: Life, the Universe and Everything in a Nutshell

In *What a Wonderful World*, Marcus Chown, physicist, broadcaster and consultant for the *New Scientist*, applies his deep understanding of complex things to simple questions about the workings of our everyday lives. Lucid, witty and hugely entertaining, the book explains the essence of our existence, stopping along the way to answer questions such as why do we breathe? How does the brain work? Why did life invent sex? Does time really exist? And how did an advanced breed of monkey like us get to dominate the Earth?

'Reading a well-written popular science book is one of the great pleasures of modern times, and this . . . affords that pleasure in abundance.' Brandon Robshaw, *Independent*

'Crammed with brain-boggling facts.' *Mirror*

'A brilliant book that makes you realise how much you don't know.' *Love Reading*

**ff**

# We Need to Talk About Kelvin:
# What Everyday Things Tell Us About the
# Universe

The reflection of your face in a window tells you that the universe at its deepest level is orchestrated by chance.

The static on a badly tuned TV screen tells you that the universe began in a big bang.

In fact, your very existence tells you this may not be the only universe but merely one among an infinity of others, stacked like the pages of a never-ending book . . .

'Perfect for someone who wants a non-intimidating intro to modern physics, or a precocious teenager who won't stop asking why.' *New Scientist*

'Chown writes with ease about some of the most brain-bending of concepts and makes you really think about science.' BBC *Focus Magazine*

'Chown writes very fluently, helping us to visualise things with matchboxes and Lego bricks.' Steven Poole, *Guardian*

ff

# Afterglow of Creation: Decoding the Message From the Beginning of Time

It's in the air around you. It carries with it a baby photo of the Universe. Its discoverers mistook it for pigeon droppings yet still won the Nobel Prize. *Afterglow of Creation* tells the story of the biggest cosmological discovery of the last hundred years: the afterglow of the big bang. The result of this find was a 'baby photo' of the universe – sensationally described as 'like seeing the face of God' – which revolutionised our picture of the cosmos. Marcus Chown goes behind the initial hysteria to provide a lively and accessible explanation of this hugely important research – and gives us a compelling and exuberant tale of the human side of science.

'Superbly captures the spirit of scientific discovery.' *Sunday Times*

'A very good piece of storytelling.' *New Scientist*

'A "science for dummies" take on creation, built around an account of the discovery of radiation ripples from the Big Bang.' *The Times*